CITIZENS AND NATION:
AN ESSAY ON HISTORY, COMMUNICATION, AND CANADA

Grandmother Andre told stories in front of a campfire. Elizabeth Goudie wrote a memoir in school scribblers. Phyllis Knight taped hours of interviews with her son. Today's families rely on television and video cameras. They are all making history.

In a different approach to that old issue, 'the Canadian identity,' Gerald Friesen links the media studies of Harold Innis to the social history of recent decades. The result is a framework for Canadian history as told by ordinary people. Friesen suggests that the common people's perceptions of time and space in what is now Canada changed with innovations in the dominant means of communication. He defines four communication-based epochs in Canadian history: the oral-traditional world of pre-contact Aboriginal people; the textual-settler household of immigrants; the print-capitalism of the nineteenth and twentieth centuries; and the screen-capitalism that has emerged in the last few decades. This analysis of communication is linked to distinctive political economies, each of which incorporates its predecessors in an increasingly complex social order.

In each epoch, using the new communication technologies, people struggled to find the political means by which they could ensure that they and their households survived and, if they were lucky, prospered. Canada is the sum of their endeavours. *Citizens and Nation* demonstrates that it is possible to find meaning in the nation's past that will interest, among others, a new, young, and multicultural reading audience.

GERALD FRIESEN is a professor in the department of history at the University of Manitoba.

CITIZENS AND NATION

An Essay on History, Communication, and Canada

GERALD FRIESEN

UNIVERSITY OF TORONTO PRESS
Toronto Buffalo London

© University of Toronto Press 2000
Toronto Buffalo London
Printed in the U.S.A.

Reprinted 2009, 2013

ISBN 0-8020-4709-2 (cloth)
ISBN 0-8020-8283-1 (paper)

Printed on acid-free paper

Canadian Cataloguing in Publication Data

Friesen, Gerald, 1943–
Citizens and nation : an essay on history,
communication, and Canada

Includes bibliographical references and index.
ISBN 0-8020-4709-2 (bound)   ISBN 0-8020-8283-1 (pbk.)

1. Canada – Civilization.   2. Canada – Social conditions.
3. Communication – Canada – History.
4. Canada – History – Philosophy.   I. Title.

FC95.F74 2000   971   C99-932871-9   F1021.F74 2000

University of Toronto Press acknowledges the financial assistance
to its publishing program of the Canada Council for the Arts and
the Ontario Arts Council.

This book has been published with the help of a grant from the
Humanities and Social Sciences Federation of Canada, using funds
provided by the Social Sciences and Humanities Research Council
of Canada.

University of Toronto Press acknowledges the financial support for
its publishing activities of the Government of Canada through the
Book Publishing Industry Development Program (BPIDP).

Canadä

For Jean, Alex, and Joe

Contents

Acknowledgments

This book began as a series of four lectures to European teachers of Canadian studies. I would like to thank Giovanni Bonanno of the University of Messina and his Italian colleagues for launching this project and for their kind hospitality; and William H. New, of the English Department at the University of British Columbia, whose views on these same subjects – now published in *Land Sliding: Imagining Space, Presence, and Power in Canadian Writing* (Toronto: University of Toronto Press, 1997) – and sharp editorial eye have contributed much to its final form.

I would also like to thank my colleagues at the University of Manitoba and the staff at the Dafoe and St Paul's College libraries; former students, including Royden Loewen, Jeff Taylor, and the late Angela Davis; the members of history seminars 438, 470, and 472; those kind souls who read drafts of chapters and even of the entire manuscript, including William Brooks, David Frank, Stephen Garton, Allan Greer, Greg Kealey, John Kendle, Ian MacPherson, Del Muise, Tom Nesmith, Paul Rutherford, Veronica Strong-Boag, Mary Vipond, and Robert Young. I presented parts of these chapters on other occasions, and I would like to thank my hosts and the students at the University of Sydney, Macquarie University, the University of Alaska at Anchorage, the University of Windsor, and the University of New Brunswick, as well as Susan Jackel and colleagues at the University of

Alberta, Jacqueline Peterson and colleagues at Washington State University, and colleagues at Trier University and the Grainau conference.

This book juxtaposes the life stories of a few citizens with the history of international communication, economics, and politics. In seeking such citizens' stories, I embarked on an unusual and rewarding odyssey that involved filmmaker Graydon McCrea of Edmonton and Alestine Andre, who is now executive director of the Gwich'in Social and Cultural Institute in Tsiigehtchic, Northwest Territories; Dan Soucoup of Nimbus Publishing, Halifax, and Joe Goudie of Labrador; Rolf Knight, who lives in Burnaby, British Columbia; and 'Roseanne,' the person who agreed to be interviewed as a representative of contemporary Canadian life. I thank them for their cooperation, their permission to use the material that follows, and their interest in this book.

Gerald Hallowell of the University of Toronto Press has provided unfailing encouragement and support during this long process. John Parry has edited the manuscript with care and creativity. To them, and to Jill McConkey and Frances Mundy and Bill Harnum and the many others at the Press who ensure that such books appear and find an audience, I would like to say a heartfelt thank you.

I dedicate this book to members of my immediate family who have endured its preparation with tact and generosity of spirit and have offered alternative views about priority and perspective sufficiently forcefully to ensure that it has finally left my desk.

Gerald Friesen
University of Manitoba
Winnipeg

CITIZENS AND NATION:
AN ESSAY ON HISTORY, COMMUNICATION, AND CANADA

Introduction

Grandmother Andre, who lived all her life in the Mackenzie River region of the North-West Territories, told stories in front of a campfire. Elizabeth Goudie, 'woman of Labrador,' laboriously wrote her own memoir in school scribblers. Phyllis Knight, a German immigrant in Vancouver, taped hours of interviews, and her son then edited them into a single volume. Works documenting the lives of Roseanne, Frank, and Simonne – representatives of today's society – make frequent mention of television and computers. These six texts, the media in which they were communicated, and the storytelling they represent testify to the structure of human history and to the character of Canada. At least, that is my contention. This book outlines how the very acts of communication – the social contexts created by voice, writing, print, and modern electronic forms – establish a framework for citizenship and nationality and thus for Canada.

We live in a moment when change is said to be inescapable, perhaps oppressive. Doomsayers suggest that even time and space, the fundamental dimensions of life, are not only under siege but have been annihilated, leaving us bewildered and without a pole star. The sense we have of ourselves – our identity – is fluid, they say, and each of us may partake of dozens of different identities in the course of a lifetime. Nations are losing influence, others claim, as power moves up to multinational and down to local institutions. Politics, according to some pessimists, has ceased to be a fruitful sphere of activity because the shifts in national and personal circumstances have left us cynical and ungovernable. This crisis – in politics, economics, culture, and communication – results from the speed of invention and the ready adaptability of today's managers, financiers, and engineers. It is made worse by ordinary citizens' inability to create the appropriate institutions through which to respond to these changes.[1]

Works of history are supposed to offer us a sense of perspective in just these circumstances. They remind us of past realities and illuminate the significance of today's novelties. In an ideal world, they also build bridges between present and past. Previous generations of Canadian historians have created widely circulated versions of a shared past. W.L. Morton's *The Canadian Identity*, a representative work of this type published in 1960, is built on the thesis that to be Canadian is, above all, to possess a particular 'political nationality.'[2] Morton selected a handful of themes which, he believed, were central to the Canadian experience: a northern and maritime setting, dependence on founding empires centred in Europe, and association with British parliamentary institutions, the monarchy, and the Commonwealth. His choices still have merit although I explain in the pages that follow why the passing of forty years has brought different themes to the fore.[3]

Narratives that purport to offer short histories of exceptionally long durations must simplify outrageously. This book may seem to claim ten thousand years of human activity as its sphere of competence, but such an ambition is clearly absurd. Rather, this book is about *today's* world. It is a history of the present. The phrase may seem unlikely, but it actually conveys a truth. What drives the following story is a quest not for all the variations in the past but for some of the most important legacies that shape contemporary Canada. Thus this narrative canvasses just a few of the forces that distinguish a modern, rich, well-situated, yet divided community.

A familiar saying declares that 'history is what people make of their geography.' R. Cole Harris, the Canadian geographer, has rephrased it: 'A society and its setting cannot be conceptualized separately.'[4] Although I agree, I would add that what is significant in 'setting' changes. The phrases 'dominant communication system' and 'perceptions of time and space,' which recur often in this book, underline my conviction that the way in which a society communicates shapes popular assumptions about how the world works.

In the following pages, I suggest that over the entire course of human history in northern North America common people have experienced four constructions of the dimensions of time and space. I suggest too that we can correlate these constructions with four dominant communication systems. The first construction of time and space, 'oral-traditional,' which we can relate to the cultures of Aboriginal peoples, commenced with their presence on this continent and endures in some major respects to this day. We should see it as one of the founding elements of Canadian as well as of human society. The second – 'textual-settler' – commenced with the partial eclipse of that oral society by a literate one, which arrived with European immigrants and evolved during the succeeding centuries. In this second phase, the common people experienced more indirect than direct contact with communication systems built on literacy and on the evolving alphabet-letterpress-print revolution. However, the implications of living within this 'textual community' expanded, spreading eventually to almost all residents of the society. The consequences of this epoch, too, remain with us. The third, which we can see as an extension or revision of the second, travelled on a variety of technological breakthroughs, including the telegraph, the daily press, film, sound recording, and radio. Because it relied on schooling, mass production, and nearly universal literacy, this construction has been labelled 'print-capitalism.' Its effects are also evident in many of the institutions that shape our lives today, including the nation of Canada itself. The fourth dom-

inant version of space and time – ours, 'screen-capitalism' –
is said to have superseded all previous cultures as a consequence of the introduction of television and computers, the
refinement of transportation and production systems, and
the development of global corporate organization as well as
of consumption-driven individual experience. Future scholars might, with the advantage of a longer perspective, sub-
divide the introduction of television and electronic data
transmission via computer, satellite, and cable into several
distinct cultural processes or, alternatively, lump them
together with the third phase of communication inventions
as a single cultural episode in the nineteenth and twentieth
centuries. However, I am proposing to treat the most recent
half-century, the days in which our own lives are unfolding,
as a fourth construction of space and time.

This is not to say that there are four sharply delineated,
exclusive, and autonomous 'cultures' in Canada's past. The
changes in the communication environment have been real
enough, but they are also cumulative. True, each reconstruction of time–space perceptions has only one point of departure: the invention of alphabet, printing press, and paper and
their arrival in this part of the world; diffusion of telegraph,
telephone, and radio; and the rapid adoption of television,
computer, and sample survey. However, each builds on its
predecessor according to the distinctive rhythms dictated by
the land and the people of this place. Each advance in communication technology therefore contributes to the reconstruction of perceptions of time and space; each supplements
and complements, but does not erase, its predecessor. There
is at any one time only one place, one evolving community
within an international communication process.

Canada's history has often been told from the vantage
point of those who possess the resources that we recognize
as power – wealth, military authority, political influence –
but this book looks at it from the other side, from the perspective of those who feel that they are responding to events
around them rather than initiating the changes. It extends

the definition of what is powerful to include access to and mastery of the leading instruments of communication in each epoch of human history (literacy in the Middle Ages, for example), as well as such conventional measures as money and guns.[5] I weave my tapestry from the history of communication and from cultural history, especially the changing perceptions of time and space, and from ordinary citizens' responses to such changes. In other words, I build my alternative history of Canada not on the French fact or the British monarchical and parliamentary inheritance, not on discussions of staple goods such as fur and wheat, and not on the stories told by privileged writers and artists. This work relies on the new methods of social and cultural history that have become so useful in the writing of Canadian history in recent decades. Nevertheless, it is connected to Canada's earlier historical writing by its concern for the expression of identity, its interest in how ideas and aspirations are conveyed within large social groups, and its concentration on this particular place. It is an essay that argues for ordinary citizens' role in shaping this nation. One central theme in the story, I suggest, is people's struggle to master changes in the dominant means of communication.

PART ONE

Oral-Traditional Societies

1

Genealogy and Economy

In the concluding pages to his landmark history, *The Fur Trade in Canada*, published in 1930, Harold Adams Innis offered a sweeping statement: 'We have not yet realized that the Indian and his culture were fundamental to the growth of Canadian institutions.'[1] These words, written by one of the country's intellectual leaders, constituted an important judgment. Although they undoubtedly rang true then, do they still sound accurate today? Surely, nowadays, we recognize the role of the First Nations in Canada, and surely we acknowledge Aboriginal people as founding peoples. After all, if human history in northern North America spans 20,000 years, then the story of Aboriginal people is the exclusive focus in 19,000 of those years; if 10,000, then in 9,000 years Aboriginal people alone are present. This very long span of occupation will have generated a rich historical tradition, one might suppose. In that very tradition, in the popular understanding among all Canadians of a rich Aboriginal past, lies the contribution of which Harold Innis wrote.

I think that Innis's observation about popular ignorance of the Aboriginal role in this country still carries weight. Despite the work of a few pioneers in scholarship and public service, and despite the tenacious stand of Aboriginal people themselves, the story of the First Nations has not been integrated into the narratives and myths of the Canadian people. As the Royal Commission on Aboriginal People declared in its 1996 report, 'We [Canadians] retain, in our conception of Canada's origins and make-up, the remnants of colonial attitudes of cultural superiority that do violence to the Aboriginal peoples to whom they are directed.'

Recent academic teaching has emphasized Aboriginal autonomy in the fur trade, Aboriginal insight into humans' relations with the natural world, and Aboriginal long suffering in the face of pain inflicted by others. Nevertheless, parallel with this supportive atmosphere in the world of print and classrooms, the daily life of many Aboriginal people in Canada has been marked by poverty and violence. Indeed, the gap between academic interpretations and contemporary Canadian awareness mirrors the widening gulf between Aboriginal and other citizens.

The argument in chapters 1 and 2 rests on the belief that Aboriginal people possess a profound sense of their continuity in this land. It suggests that their convictions about their location in this world underlie all Canadian culture. This chapter introduces a memorable documentary film on Aboriginal life and then explores two factors revealed there – genealogy and a distinctive Aboriginal traditional economy – that demonstrate the depth of connection between Aboriginal people and this land. However, it concludes that these two elements are not sufficient to account for Innis's judgment that Aboriginal peoples are fundamental to Canadian institutions. Chapter 2 turns to story-telling, the traditional dimensions of time and space, and Aboriginal approaches to politics. It proposes that we can best understand the authority of the Aboriginal heritage in today's Canada – the reason for its central role in contemporary life – by looking at such cultural and political factors.

I believe that Aboriginal nations should occupy 'a place of honour in our shared history,'[2] as the members of the royal commission declared. I believe too that all Canadians should understand the elements of a common narrative embracing both Aboriginal and non-Aboriginal Canadians. As Harold Innis recognized, the issue of Aboriginal continuity is central to the construction of a Canadian historical narrative.[3] These two chapters assert that there was in the past and there remains today an element in Canadian life that is Aboriginal in character. The task of all Canadians, I suggest, is to understand what this means.

I

Graydon McCrea's *Summer of the Loucheux: Portrait of a Northern Indian Family* is a balanced, instructive illustration of the place of Aboriginal people in Canadian history and society.[4] This brief documentary film, released in 1983, spends a few days with four generations of a Gwich'in (Dene) family at their camps on the Mackenzie and Tree rivers in the subarctic lands of northern North America. Its subjects include Alestine Andre, then twenty-eight; her sixty-nine-year-old father, Hyacinthe; her grandmother, Julienne; and her three-year-old niece, Lisa. They fish at these sites during July and August, just as their ancestors did in generations past. From their ancestors, they have learned the best techniques for filleting the white fish, drying them in the wind and sun, and then smoking them to preserve them over the winter. The wealth of knowledge necessary to do this work – about the kind of wood to burn in the fire, where to set the net, how to construct the smokehouse, how to pack the fish – is as precise and subtle as any other type of knowledge that has to be communicated from one generation to another. The theme of the film is the transfer of such community knowledge – the process of 'cultural reproduction.'

McCrea's documentary focuses on the fishing camp and moves with family members as they prepare nets, travel in the motorboat, land and fillet the catch, and then smoke the fillets. Everyone has a role in the camp's activity. Hyacinthe handles the boat and the fishing, the women supervise daily food preparation – although the roles are interchangeable – and little Lisa learns by watching and imitating. Alestine has recently commented that she has since become familiar with

The Andre family fishing camp on the Mackenzie River, 1979.

most of these skills and that each August she still carries out these tasks.

Summer of the Loucheux makes it very clear that the natural environment framed much of the experience of the Andre family. Alestine explains that she had travelled and harvested resources with her parents throughout her childhood. She says at one point: 'I rely a lot on my father ... [He] spent almost all his life out on the land.' At another moment, she recalls standing at her mother's fish table, just as Lisa is doing now. She also notes without irony that adaptation to a residential school miles from home had been hard for her but that life in the fishing camp presented 'no worries.' She describes the summer fishing as a way to re-energize and to escape from the worries of town life. She reports that only two of her eleven siblings regularly visit the fishing camp, but she hopes that Lisa will 'catch on to wanting to come out here.' She says quietly that 'the land means a lot' and that her people rely on it. The viewer observes that the Andre family lives comfortably in this tent at the river's edge and that,

Four generations of the Andre family (from left): Marka Bullock
(Alestine's aunt), Julienne Andre (Alestine's grandmother),
Alestine (with sunglasses), Lisa (Alestine's niece, with rabbit),
and Hyacinthe Andre (Alestine's father).

from the evidence, their ancestors followed similar practices
over the centuries.

The film provides a wonderful portrayal of the easy pace
of camp life, with less talking and more doing in daily activity than my students, at least, are accustomed to. Some who
have watched the documentary have been moved to say that
nothing happens in it or that it is slow to make its case. But
therein lies a strength. After all, we are seeing how one generation communicates with the next, and in this family
actions often take the place of words. In classroom discussions, students mention the Andres' respect for tradition and
the evidence of continuity from a distant past to the present.
They recognize that at the film's heart lies a gentle illustration of the ways in which a group's survival skills and sense
of life's meaning are passed on to those who follow.

This short, quiet film offers enough to construct a history of Aboriginal people in northern North America. Men and women worked together to carry out the necessary daily tasks. The family travelled purposefully – not aimlessly – from one resource zone to another as the seasons changed. Although the Andre family's fishing camp contained metal knives, outboard motors, chainsaws, and a radio, it did not operate at the direction of European Canadians. Rather, one could describe the two worlds as interdependent. Nevertheless, the larger transition to European economic rules in the nineteenth and twentieth centuries affected the fishing camp as it also changed the world. The negotiation of a treaty between local citizens and the Canadian government, as well as the crucial policy decisions associated with the Indian Act, also shaped the Dene experience. Alestine mentions the problems that she experienced in an Aboriginal residential school far from her parents. She expresses unease about the welfare payments in her northern community. As the story closes, a huge motorized barge chugs down the river, its noisy engine drowning out all other sounds in the camp.

Summer of the Loucheux constitutes in effect a survey of Aboriginal life in the present and the past. It introduces the themes that we must consider if we are to understand Innis's sweeping statement about Aboriginal people's being fundamental to the growth of Canadian institutions.

II

Consider the matter of genealogy. Today's Aboriginal people are the direct descendants of families that inhabited this land many centuries ago. Can one find in this genealogical connection what Innis was looking for – an explanation of why Aboriginal people are fundamental to Canadian institutions? In a world where numbers are inevitably used to estimate influence, the size of the Aboriginal population must affect the power that it can exercise. And that size is relatively small. We will probably never know the Aboriginal population of northern North America in the seventeenth century, but it was much larger in the year 1600 than it was three centuries later, by which time contact with new diseases and a different social and political order had taken its toll. At the beginning of the twentieth century, under the Indian Act, the Canadian government enumerated only 100,000 Indians. Today, there are about 600,000 registered Indians.[5] We can include as well many people of mixed European and Aboriginal descent – perhaps as many as a half million – whom the constitution now categorizes as Métis.[6] In addition, there are nearly 40,000 Inuit. Each of these three groups – 'Indians,' 'Métis,' and 'Inuit' – are named in the Constitution Act of 1982 as possessors of a distinct political identity within the general category of Canadian citizenship.[7] In sum, more than one million Canadians, or just under 4 per cent of the total population, can trace their heritage back to Aboriginal inhabitants who have dwelt on this land for centuries. This is a small community on which to assert a fundamental role for the First Nations in the development of Canadian institutions.

Canadian school children usually learn that Aboriginal people reached the Americas from Asia and have lived in North and South America for at least twelve thousand years, and perhaps thousands more. In universities, students hear that Aboriginal people have lived in the Americas from 'time immemorial.' In both approaches, classes are taught that the First Nations spoke many separate languages and developed very different economies and cultures as their migrations carried them into new territories and offered them new resources. Thus, whether thinking in the relatively simple terms of Asian migration or in the more complicated concepts associated with time immemorial, Canadians do acknowledge the connection between today's First Nations and the original inhabitants of this hemisphere.[8]

There is, however, a grudging quality to Canadians' recognition of the Aboriginal heritage – a reluctance to grant exceptional status to peoples who have moved many times over the centuries. And indeed it is true that migration and change have been crucial in Aboriginal history. It is certain, for example, that a number of band-based societies changed their locations during the three centuries between 1600 and 1900. The westernmost Cree, for example, lived near Hudson Bay in the 1640s, near Lake Winnipeg in the 1740s, and in the North Saskatchewan River valley as far west as the Rocky Mountains in the 1840s – a total shift of over two thousand kilometres.[9] Major cultural changes accompanied the changes in homeland. While some of the Cree continued to live in forests and along the forest-parkland margin, others were attracted by the life of the plains and the buffalo hunt. Within a few years in the late eighteenth century, these latter groups adopted the horse and the war raid and the material culture of their neighbours on the prairies.[10]

Such stories of Aboriginal movement sound familiar to most Canadians. They are, on the surface, the mirror image of European-, African-, and Asian-Canadian family sagas. Homelands across an ocean, immigration to this new land, adaptation to another environment, moves from one part of the continent to another: each of these themes strikes a familiar chord. The sense of recognition was clearly expressed by a prominent Canadian politician in the 1970s (the mayor of Calgary, Rod Sykes), who declared that all Canadians, whether Native or European in origin, were 'immigrants.' The only difference, he said (thereby creating one of those images that can live in the public mind for decades), was the date on their tickets. His neat phrase conjured up strong folk memories and evoked popular support; but at its heart it was profoundly misleading.

Aboriginal Canadians are not immigrants. The difference between a history of three or four hundred years and a past of twelve thousand (or twenty or forty thousand) years is not just a matter of degree. A heritage so deep that it is measured in hundreds of generations is not comparable in any meaningful sense to the European migrations to the Americas. As a consequence, Aboriginal continuity in this place has been acknowledged in the recent codification of Canadian citizenship. Aboriginal people have a different political and legal status in the Constitution Act of 1982 because of their occupation of this land from time immemorial. And yet, although it is the first and inescapable truth about their connection with this land, it is not sufficient to convince many of the other 96 per cent of Canadians that their own history is inextricably intertwined with that of their Aboriginal compatriots. Genealogy is necessary but not sufficient to sustain Innis's thesis about the centrality of Aboriginal contributions to Canada's institutions.

III

Let us turn from time immemorial to the oft-heard assertion that there is a distinctive Aboriginal approach to economic activity. To argue that a traditional Aboriginal economy, and the thinking derived from it, are central to Canada's economic institutions today might seem silly. Canada is a technologically advanced, globally competitive capitalist community. It is not, whatever the abiding strength of earlier worlds, a traditional, family-based, subsistence society. True, Aboriginal people adapted to the environment of northern North America during thousands of years of occupation. But today's Canada offers little evidence of the economic principles and practices that they developed. Could Innis actually have believed that First Nations were fundamental to the growth of today's economic institutions?

There is no academic consensus concerning the nature of Aboriginal economic arrangements before Europeans arrived. Many early European observers used uncomplimentary terms to describe Aboriginal economic behaviour. Although they reported that Indians were starving, or that members of a band said that they were starving, these same Europeans very rarely recorded death by malnutrition among Aboriginal groups. Indeed, starvation seems to have been a relative term in the language of the hunting bands.

There were very hard times when individuals died because of food shortages, but such crises were probably more rare in North America than in Europe. Instead, what one should understand by these references is periodic Aboriginal hunger, not starvation.[11]

Perhaps Innis was thinking not of what Canada actually was in 1930 but of what it might become. Research on a wide range of hunting and gathering peoples, as reported by Marshall Sahlins, suggests that in many parts of the world such groups were able to obtain adequate food supplies with an output of two to three hours of labour each day. This, said Sahlins, constituted the 'zen road to affluence,' in which hunting and gathering peoples, by minimizing their demands for material possessions, actually enjoyed far more leisure time than do today's working people.[12] Did Innis imagine that a utopia might be built in northern North America if only Canadians returned to the conservation practices and moral economy of traditional societies?

Sahlins-like reports of a work-free, food-filled Aboriginal utopia did not come from northern North America. When Adrian Tanner lived for eighteen months with Cree hunting bands near Lake Mistassini, Quebec, in the late 1960s, he estimated that most of the group's winter food – over 70 per cent – was harvested from the land, but only with great expenditures of effort. Tanner discovered that life in the bush required endurance and involved periods of privation. There was always enough food to maintain the band and in some bountiful winters to sustain larger, multi-family groups in festive plenty. But Tanner added a warning:

> Let me stress that I am not arguing here for the view that sub-arctic hunting is highly productive for a minimum input of labour time; on the contrary, all adults are involved in long hours of often back-breaking labour to achieve a level of subsistence which the Mistassini themselves find seldom much more than adequate. Neither am I suggesting that the planned approach to hunting and trapping could continue during con-

ditions caused by a seriously impoverished environment. We have evidence of starvation at Mistassini, with deaths for example in 1834–5 and 1837–8; other deaths appear in our genealogies during the period 1915–25 ... At such times the group became dependent on small mammals, small birds and fish, which was usually the woman's productive speciality; the men scoured the countryside with a gun, living off small game, and sometime later seeking help. Hunting groups and even families were thus broken apart temporarily by starvation.[13]

No one should presume that the quest for natural food in northern Quebec or Saskatchewan much resembles the situation in Central America or the South Pacific. Food gathering was far more taxing in most of northern North America; it was not, however, a constant gamble with death.[14] Tanner's careful compilations of the monthly kill of a band in northern Quebec demonstrate how the attentions of the hunters shifted with resource availability. Beaver were taken in late winter, from January to April, as were ptarmigan and grouse; fish and waterfowl were more important in fall and spring; big game, especially moose and caribou, were hunted whenever they appeared in the district between November and May. In the nine months that Tanner spent with this band of seven adults and six children during 1967–8, the group harvested eight thousand pounds of edible meat, or about twenty-nine pounds each day. It will be apparent that Tanner witnessed an economic enterprise based on specialized knowledge.[15] Still, this economy was far removed from modern electronic marvels, and we cannot usefully compare its production with modern capabilities.

Another tack in my quest for Innis's state of mind is to emphasize the close connection between human and natural spheres in the Aboriginal universe. Did Innis have in mind the alienation of modern labourers from their places of living and working? Was he following Karl Marx in imagining a reknitting of humanity and natural world, and was he projecting a reconstruction of Canadian institutions in which Aboriginal economic practice would serve as a guide?

This argument will be familiar to Canadian school children, who are regularly told about the wonders of Aboriginal material goods. And these stories are true. The effectiveness of items in the Aboriginal domestic economy varied with the task and the raw material, but tents of caribou skin were light and warm; toboggans, snowshoes, and canoes were ideal instruments for the roles that they performed; and much of the skin clothing was light and well-adapted to the climate. However, just to maintain the balance, recall that the metal products of the Europeans – knife, axe, kettle, needle – were often more efficient than Native equivalents of stone, skin, wood, and bone. There should be no surprise in a tally that finds items of superior design in both the European and the Aboriginal domestic economy. It is worth remembering that the basic material needs of the community – food, clothing, shelter – were certainly attainable with the use of Aboriginal material goods. And all the items were easily carried by the hunting band. But none of this suggests that Innis wanted to return to an earlier technological era. If we are to discover what he was thinking about, we must look for something more basic than material goods and technology.

Innis's insight may be related to the communion with nature that he witnessed and experienced in his travels throughout northern Canada. Most Canadian youngsters have heard some version of a discussion about Aboriginal technology, but not as much about an Aboriginal science of the land. Their teachers, however, have attended debates about whether the First Nations were conservers of nature's resources or instead were as likely to despoil nature as any modern profit-taker. In such discussions, some scholars have used fur traders' accounts of wasteful Aboriginal harvests to question whether broad generalizations about the Aboriginal status as conservationists can be sustained.[16] Others cite reports of Aboriginal respect for the natural world, especially for living creatures, to contradict the suggestion that environmental carelessness is endemic to every culture, including the Aboriginal. An illustration comes from a story told by Hugh Brody, writer and filmmaker, about a hunt in British Columbia in the 1970s. A pregnant moose was shot on an icy afternoon that threatened to turn dangerously cold and on land that was probably a rancher's private property – a fact that might raise difficult legal questions for the Indians: 'The moose was butchered hurriedly but properly. As always, virtually every part of the animal was taken: the digestive system was cleaned out, the intestines separated from the other entrails, the liver and heart carefully set aside. Despite the cold, and whatever uneasy feelings were inspired by the thought that others might say we were trespassing, the hunters observed the rules and conventions that govern the butchering of a moose. On this occasion they observed also the conventions that surround the treatment of an unborn calf ... Only when all the remains were entirely hidden away was the butchering finished. Such respectful treatment of the foetus was as automatic as any other part of the butchering. Even if it did cause some delay, there was no question of its not being carried out.'[17]

Narrators of such stories often imply that Aboriginal people had a closer connection to the world of animals and plants than do other Canadians today. The most vigorous of such exponents add yet another element that troubles the empirically minded: this extra dimension is the apparently magical or telepathic process by which a hunter foresees the location and kill of game animals, especially in moments of famine. Yet this phenomenon has often been recorded by observers of Aboriginal hunting bands. A young American, John Tanner, who was captured and then adopted by an Ojibwa band in the 1790s, described an episode in which, at a moment when their camp faced great hunger, an old Aboriginal woman foretold the location and kill of game. The men followed her directions, found a bear, and killed it, and the group was saved. Tanner records several other examples of such foreshadowing in his memoir.[18] George Nelson, a trader in the Canadian northwest in the early nineteenth century, experienced a similar augury; Nelson reported the success of the hunt and added defensively that his readers might not believe his story, but it was none the less true.[19]

The importance of being attuned to the natural world was so obvious and well-known in hunting societies that Louis Riel, Canada's most famous Métis leader, during his speech at his 1885 trial for treason, used it as an offhand illustration and amplification:

> It is not to be supposed that the half-breeds [would have] acknowledged me as a prophet if they had not seen that I could see something [–] into the future. If I am blessed without measure[,] I can see something [–] into the future ... [Riel then turned in the prisoner's box to attack his cousin Charles Nolin, who had given evidence against him] ... Nolin knows that among his nationality, which is mine, he knows that the half-breeds as hunters can foretell many things, perhaps some of you have a special knowledge of it. I have seen half-breeds who say, my hand is shaking, this part of my hand is shaking you will see such a thing to-day, and it happens. Others will say I feel the flesh on my leg move in such a way, it is a sign of such a thing, and it happens. There are men who know that I speak right.[20]

At crucial times, when significant changes in group activity or real crises had to be addressed, dreams or forecasts often helped the Aboriginal people to make choices. Adrian Tanner concluded from his experience among the Mistassini Cree in the 1960s that such divination 'was conducted as an intellectual exercise which accompanied the collection of hunting data ... The kind of information divination gives,' he writes, 'is just that kind that cannot be known in advance from an examination of environmental signs. Divination fills in gaps in knowledge which cannot be learned from the environment.'[21]

Relations between human and animal in the world of

hunters were much closer than most contemporary Canadians would acknowledge or even imagine. The link was uncertain but it was also subject to a degree of imagination that demonstrated to the Aboriginals that animals could be generous to humans and that humans could be very resilient in the face of challenges to survival. Wise, good, strong hunters (strong in the sense that they possessed both skill and personal authority) would triumph over adversity precisely because they could become attuned to the animals' world – the spiritual world, an Aboriginal hunter would say – and could then imagine or foresee or will the location of food. Was this what Innis was thinking about? I suspect that there is something in this suggestion, although it will require some very careful consideration to substantiate.

The pages above have looked at the question of an Aboriginal foundation for modern Canada in economic terms. I have discussed relations between Aboriginal people and the environment as if they dealt simply with the gathering of food and the maintenance of material life. But surely this economic analysis must break down when confronted by the modern economic system. It makes sense to assume that, before the arrival of Europeans in the seventeenth century, Aboriginal peoples had developed an appropriately adapted economy and an effective understanding or science of the land. But what happened thereafter? Does the continuity of economic practice established during thousands of years then break down? I suggest that the answer is 'yes' and that Innis would agree.

Europe's evolving economic assumptions, along with its merchants and its products, were carried around the globe after 1400. World trade was revolutionized by Vasco Da Gama, Christopher Columbus, and Jacques Cartier, by cotton and steam and telegraphy, by joint stock companies and imperial civil servants. Most important of all, the European economic system changed in character during and after the Industrial Revolution, and, willy-nilly, so did the economy of northern North America. No longer based simply on

trade and exchange, the capitalist system of economic organization soon extended around the world. Aboriginal people
were swept into this global economic revolution. In the process, land and labour themselves became commodities. And
other means of production, including many aspects of
knowledge, increasingly were converted into property and
privatized.

When the twentieth century opened, observers were laud-
ing or denouncing the new European system of economic
organization, which they called 'capitalism,' and celebrating
the world's imminent economic integration. Given western
Europe's trade conquests during these centuries, and especially the eventual collapse in the late 1980s of the chief competing system in central and eastern Europe and China, this
span of five centuries ends with the seeming victory of the
capitalist way. How could there be an Aboriginal cast to
modern Canadian economic institutions?

The economic system in northern North America during
the centuries of transition from traditional to capitalist institutions – the system in which both Aboriginal and European-
Canadian commodity-producing households participated –
was not capitalist in today's sense: it was based instead on
family production units. Its economy was bound by and
rooted in broader social concerns. Residents, Aboriginal or
not, knew as a matter of faith that a needy household could
have access without the permission of 'private owners' to
such means of production as fishing grounds, cultivable
fields, and animal runs. They presumed access, as well
(because any skilled family could make them), to tools such
as nets and snares necessary to harvest fish, crops, and animals. Continuity in subsistence activities established the conventions of life – or of 'economic life,' as one might put it
today, in our more compartmentalized society.[22]

The characteristics of this not-quite-capitalist mode of production will explain why many recent students of fur trade
history have been so preoccupied with the theme of Aboriginal independence. The trend in this scholarship has been to

emphasize the control, self-respect, diplomacy, and rational economic calculations of Aboriginal people in the era of the fur trade. This writing has also proposed Aboriginal reasons, not explanations derived from European actions, for migration and culture change. The acuity of Aboriginal leaders often receives special comment. The authors of these works have concluded that many Aboriginal communities flourished outside the calculations of profit and loss maintained by Europeans.[23]

As the decades passed, however, the First Nations were drawn ever more tightly into the world economic system. Trade with private fur companies helped to increase pressure on resources and to reduce Aboriginal peoples' ability to deal with scarcity. The consequence, as A.J. Ray has argued in the case of the dominant fur enterprise – the Hudson's Bay Company – was the transfer of 'social costs' to the Canadian government, while the company continued to reap profits from the exchange of goods for furs. Ray concludes that these social or welfare costs were 'deeply rooted in the fur trade' and were 'a necessary by-product of several processes: economic specialization by native peoples, a concomitant decreasing spatial mobility, European control of food surpluses and the depletion of resources. Reinforcing these were the labour policies, wage schedules, and standards of trade that assured the Hudson's Bay Company large gross profit margins in good years under near monopoly conditions, but that also allowed native peoples only a marginal return.'[24]

Aboriginal people necessarily made the transition to the capitalist way in economic matters, whatever their dissatisfaction with the culture of the newcomers.[25] Their economic strategies were sometimes inadequate, however, as in prairie Canada in the 1880s during the shocking food shortages caused by the near-extinction of the buffalo. Nevertheless, Aboriginal households adapted. In Manitoba, for example, the Dakota chose several paths, including wage labour and gardening and commercial agriculture, to sustain their communities between the 1880s and the 1930s, when they encountered renewed crisis because they did not have the political power to implement further adaptive strategies. Their minimal capital resources, limited educational opportunities, and meagre land base then propelled some First Nations groups into an impoverished existence on the margin of society.[26]

Such a rough sketch of European and Aboriginal economic history conceals more than it explains. It offers a triumphalist view of European conquests and shallow assumptions about Aboriginal defeat. I introduce just two points of debate. Recall, first, that there are many varieties of capitalism. Note the differences among France, Sweden, and the United States, for example, not to mention China, India, and Japan. 'Europe's' capitalist way is not some single, global norm. Second, the great advantage of this modern capitalist system is its ability to establish means by which vastly different societies and economies can communicate in the common currency of the market-place. But this capacity does not eliminate social and cultural differences, including those in systems of work discipline, concepts of status, and preferences for goods. As Marshall Sahlins has written, 'The world-system [an economic network] is the rational expression of relative cultural logics, that is in the terms of exchange-value. A system of cultural differences organized as a division of labour, it is a global market in human frailties ... A history of the world system, therefore, must discover the culture mystified in the capitalism.'[27]

Sahlins is attempting to understand the crucial distinction between the capitalist system's integrative or homogenizing power and a local culture's ability to resist such encroachments. This is an important point, which should be respected, because it rests on the admirable political principle of valuing diversity in a community. However, what does such a discussion say to Innis's assertion? These two issues of debate – variety within the world economic order and the crucial role of cultural difference within capitalism – might seem to restore the possibility that Innis was hoping for an Aboriginal victory in the struggle to define Canadian economic institutions. They might even seem to suggest that First Nations adaptation to European-Canadian capitalism never erased the distinctive economic principles of First Nations people. Such an argument, as I noted above, might be built on an alleged Aboriginal concern for conservation, or an alleged Aboriginal selflessness in relation to property and resource sharing (such as that described by scholars as a 'moral economy'), or on some other set of characteristics. Is there evident in Canada today a distinctive Aboriginal twist to national economic institutions?

I am not persuaded by this argument. Nor, I suspect, was Innis. It is true that capitalism came late to northern North America. Indeed, like other peripheral areas of the world economy, as is apparent in parts II and III of this book, Canada did not become immersed in the global trading system until the nineteenth or, in some corners of this land, until the twentieth century. It is also true that during the early 1980s Alestine and Hyacinthe still found niches in their society where they could practise age-old activities for a few weeks each summer. Moreover, for several hundred years after initial contact with Europeans, Aboriginal people were able to learn the ways of the new order, to adapt gradually to markets and profits and debts and weights and measures, and to retain a measure of autonomy. Adapt they did, however. The traditional Aboriginal approach to production and exchange does not now constitute a dominant perspective in the contemporary Canadian economy. A distinctive Aboriginal approach to property and production ceased to be dominant some time ago. Instead, the assumptions of European capitalism – markets, property, and the rest – have become the prevailing rules of Canadian society. Aboriginal perspectives have been relegated to the sidelines, where dissenting individuals and a number of communities, like that of Alestine's parents, can preserve and pass them on. The Aboriginal view became not so much an economic principle, from the perspective of the larger society, as a cultural survival.

Harold Innis's dictum about the centrality of Aboriginal culture in Canada's institutions cannot rest on genealogy alone. Nor can it rely on an alleged Aboriginal influence in the dominant economic institutions of contemporary Canada. Rather, one must turn from economics to culture to find why Aboriginal people are 'fundamental to Canadian institutions.'

2

Interpreting Aboriginal Cultures

If Aboriginal views do not underlie Canada's economic institutions, and if Aboriginal genealogies do not convince today's Canadians of their country's indebtedness to Aboriginal precursors, how are Aboriginal people 'fundamental to the growth of Canadian institutions'? The answer, as we see in this chapter, lies in two related realms – in their cultures and in their politics. First Nations people once experienced the basic dimensions of life, the dimensions of time and space, in terms of their relations with the natural world. They did so in specific, identifiable places in northern North America. The dominant mode of communication in their communities was speech. What is more, their political commitment, evident in hundreds of episodes of community solidarity and resistance, has ensured that their knowledge of their group's distinctiveness – their connection to place – lies at the heart of today's Canada.

I

Summer of the Loucheux illustrates this Aboriginal sense of continuity. The climax of the film occurs when the Andre family travels across the river to visit the fishing camp where Alestine's grandmother, ninety-three years old, is living. Alestine says that she had always known that her grandmother was a very special person and that 'we had to prepare a place for her. That's one of the important things that I learned from my grandma – that you always look after your old people.' The very old woman was uncertain about whether to participate in the film project. However, she decided to do so and prepared carefully for the arrival of family and film crew, to the extent that she clothed herself in a striking rabbitskin coat that she had woven years earlier. When all was ready for the filming, she lit her pipe, leaned forward, and began to speak in Dene dialect. In voice over, a narrator reads Hyacinthe's translation of her speech:

> I was born at the mouth of Travier River. I grew up in the country with Sandy River to the east and Tree River to the west, so I was born in the middle. That's why I think of all this country as my own. It's just mine.
>
> Now I am going to tell a story about up the Arctic Red River [Tsiigehtchic]. Nobody stayed in towns then. By this time in the summer, the people had left for the bush already. From the mouth of the Arctic Red River, you can paddle less than a mile. From there, you have to tie a rope to your canoe. Then you walk along and pull the canoe. That's the way we travelled, all the way up the river, for a long way.
>
> Now we leave the canoe, we are heading for the mountains. Lots of mosquitoes. We take no white man's food. Sometimes we've got nothing to eat.
>
> Now we come to a fish lake. We set nets on that lake and get a few fish. When we get to the mountains, we see mountain sheep – way on the top of the mountain. The sheep are coming down – right to the foot of the mountain. They look just like rabbits. We hunt the mountain sheep there until early fall. And when we finish we soon begin to see lots of caribou in the high meadows. We are living on good food then. Why would we stay in town?
>
> When we've got enough dry meat, we go back to the forest to trap for the winter. Now there's lots of snow. We've got caribou skins to live in. We build a big lodge. We get fur and we go to the fish lake. And the women stay there.

The men go to Fort MacPherson for some supplies. When
the men come back, we start to move again, to the mountains.
The men go first. They go on snowshoes just with an axe and a
gun. The women load up the sled, follow the men, and look
for the axe. When you find your husband's axe, that's where
you set up your camp.

The children are crying. They're so cold. You pack the children next to you. I don't know how many times I pack your
father like that. That much I looked after him. And today my
own boy, he looks after me like a baby.

I like to tell old stories about all those good people and
what they're doing. I talk about it, but, right away, I just feel
full of tears in my heart.[1]

Grandmother's story is effective as drama. It is also an in-
valuable historical document. Students watching the film
respond to the old woman and listen carefully to her tale.
They readily recognize that she illustrates the place of history and narrative in a community. And this is an important
insight. To grasp the significance of the old woman and her
story is to grasp the implications of Harold Innis's statement
about the Aboriginal place in Canadian life. But how to articulate and communicate this lesson?

The film's achievement is that it captures the feeling of continuity – the sense of historical connectedness – embedded in
these individuals. This structure of feeling is, I suspect, what
attracted filmmaker McCrea to Alestine's story.[2] And it is
undoubtedly what prompted him to place grandmother's
narrative at the climax of his work. Although her story seems
to lack drama or action, it is powerful because it conveys
authority. It constitutes the assertion of an Aboriginal sense
of self and sense of place. It also communicates a message of
hope: Grandma is telling her son, granddaughter, and great-
granddaughter that their people survived in the past,
through good times and bad; that there have always been
times just like the moment they are enjoying as she speaks,
when their people feel at ease in the land, telling stories as the

sun sets and the fire burns low on a warm summer's evening; and, just as these times have been rekindled today, so shall they occur again. Alestine comments when speaking of her grandmother: 'By the stories that are being told ... it gives you a sense of pride that there were times back in the old days not only up in this area but other parts of Canada where Indian people were very independent.'

Let us examine grandmother's reflections on her life, using time and space and then politics as entry points to the structures of feeling contained therein.

II

Time and space are complicated notions. In today's world, we divide time into minutes, days, years, and centuries, and we often experience it as an insistent, linear, and even monetized unit of life's dimensions. In the past century or so, developments in scientific analysis have extended the range of human comprehension of time as far back as an event billions of years ago, the 'Big Bang,' from which moment the evolution of the universe is said to have begun. Such intellectual breakthroughs may not be a regular part of daily conversation, but they have established for all of us an extraordinary vista of measured moments from 'the beginning of time' to the present. Is it any wonder that, in the modern age, time seems to be a fixed measuring rod of not quite infinite proportions?[3]

Space is, in contrast, increasingly vague. We all know that it can be measured in kilometres and cubic centimetres, described in terms of shapes, sold by volume and area, mapped by distance and nationality, and that, in each of these contexts, it is a concrete and seemingly objective thing. But we also know that, in the last several centuries, space has been made to seem a less definitive, and even less important, dimension of life. After all, from the steam-engine and telegraph to television, satellites, and e-mail, the world's inventors have enabled the rest of us to transcend the limitations of space. All of which is to say that, despite their seemingly determinate characteristics in our own age, time and space vary a great deal depending on the society and epoch in which they are experienced.

It is true that our everyday calculations of time seem to build on natural occurrences. As Anthony Aveni, a student of anthropology and astronomy, has written: 'Rare is the element in any of the calendars ... that does not grow out of the repeatable phenomena of nature's cycles, both physical and biological.'[4] Many traditional societies, for example, organized their economic activities around an *ecological time* that coincides with the solar and lunar cycles and the seasons. They also employed a second measure, *structural time*, that marks the passing of the generations and explains the interaction of groups within them. Aveni explains: 'These people do not believe in history the way we do, though they do have a sense of history. As in the events and relationships that comprise tribal life, there is a kind of immediacy to both cyclic ecological time and linear structural time among these tribal societies. Ultimate origins do not matter in their temporal time schemes. Interaction, with either nature or other people, is the real reason to keep time; and when things cease to interact or before they ever had interacted, there is no need of reckoning it.'[5]

These societies placed great store on continuity and duration. The Mayas, whose leaders were guided by sophisticated astronomical observers in their struggle to maintain dynastic power, 'did not segregate past from future as we do. For them, the past could and, indeed, did repeat itself. If you paid close enough attention to time, you could see that the past already contained the future.'[6]

This may seem to be a paradox. How can the past contain the future? Given the pace of change, and in the light of my argument above about the irreversibility of the history of communication, is it not simply wrong? On the contrary, I suggest that this remarkable Aboriginal perspective does deserve close attention. A Canadian illustration will clarify

the point. In the winter of 1994–5, an Aboriginal man resid-
ing on a reserve in northwestern Ontario saved a life in
remarkable circumstances. The Ojibwa man had been restless the previous night, bothered by a dream in which a hand
had appeared out of a lake or out of an ice-bound landscape.
In the morning, he made plans to drive on the new winter
road, a route ploughed across the lake, from Pikangikum to
the laundromat in Red Lake. He was impatient, urging his
companion to hurry up, and finally they departed. Their
truck was travelling fairly rapidly across the great broad
sweep of newly scraped road when he spotted a hand stick-
ing out of the ice. He immediately called on his friend to
stop; they skidded some distance and turned around, raced
back, and were in time to pull a grader operator to safety.
The machine had been maintaining the winter road when it
plunged through the ice to the bottom of the lake. Its operator had managed to escape from the cab but was near
exhaustion, immersed in shattered ice and freezing water,
when he was dragged to safety. The Aboriginal rescuer, having foreseen the crisis in his dream, had been able to respond
in a timely fashion. And he now had a story to tell.

How should we understand these events? The man's
dream foretold the event. The fact of the foretelling, along
with the drama of the rescue itself, made the story worth
repeating. That is, what became history – worthy of telling as
a story – had been experienced twice, in dream and in reality,
or in future and in past. History in this sense was what had
been foretold, a fact that both emphasized the importance of
the event and made its telling worthy of contemplation by
others. In this story-telling process, one's sense of time is
subject to revision. Time is less fixed, less rigid and unbend-
ing, to the Aboriginal storyteller and to those who lend an
ear to his or her narrative than it might seem to observers
unaccustomed to this perspective or unsympathetic to such
notions.[7]

Traditional approaches to time have often been recorded in
the history of northern North America. Like other communi-

ties that have relied on nature's rhythms, Aboriginal people have used natural changes – days and seasons – to measure ecological time. A northern hunting band's annual calendar might include two major seasons, winter and summer, with four additional subdivisions for early and late spring and fall; the biggest distinction lay between the time of cold and the time of warmth because of the differences in diet, transport, social organization, and band location in those two periods. The Huron names for the time units that were the equivalents of months, for example, distinguished 'days growing longer after winter, water beginning to flow in spring ... fish running, berries flowering and becoming fruit, crops maturing, birds returning, deer yarding and bears giving birth.'[8]

The Blackfoot 'winter count' – a kind of community chronicle – built on the longer rhythms of the annual cycle of winter and summer to record a structural time representing a chronology that extended over several generations, each 'year' associated with a memorable event.[9] Indeed, as Aveni observed in the case of a Dakota Sioux equivalent, 'Absolute time was marked not by the date or the number in a cycle, but rather by the most significant event occurring between consecutive winters. They painted each event in the form of a sequence of picture writings on buffalo hide: the inundation, the war, the great snowfall, the disastrous hunt, the construction of a new town by the whites, and so on. These major events constituted the Sioux's winter count – a year calendar not unlike our month calendar. In the Dakota Sioux's calendar, time and history appear to be the same.'[10]

Although these versions of ecological and structural time will seem entirely familiar, it is none the less the case that Huron and Blackfoot versions of time are not like the dimensions in which Canadians live today. Rather, they constitute a quite different notion of time itself. And it is not easy to perceive why or how the divergence in perspective occurs. Claude Lévi-Strauss, the French anthropologist, once argued that traditional peoples inhabited a 'timelessness';[11] Calvin

Martin, the American historian, wrote of Aboriginal peoples' 'astounding ability to annul time, their remarkable capacity to repudiate systematically time and history.'[12] Such metaphors require some thought.

Perhaps the best way to appreciate the divergence is to see it in action. Robin Ridington, an anthropologist, has illustrated such abstractions by referring to stories that he was told by members of a Beaver (Dunne-za) Indian community in northeastern British Columbia, with whom he lived in the 1960s. One of his contacts there, a man named Johnny, was willing to translate the last stories of his father, Japasa, for Ridington: 'My dad said that when he was a boy, about nine years old, he went into the bush alone. He was lost from his people. In the night it rained. He was cold and wet from the rain, but in the morning he found himself warm and dry. A pair of silver foxes had come and protected him. After that, the foxes kept him and looked after him. He stayed with them and they protected him. Those foxes had three pups. The male and female foxes brought food for the pups. They brought food for my dad too. They looked after him as if they were all the same. Those foxes wore clothes like people. My dad said he could understand their language. He said they taught him a song.'

Noting that when he visited this community Japasa was near death, Ridington explained that he recorded the subsequent events with care:

Japasa began to sing. The song seemed to be part of his story. It must have been the song the foxes gave him. It must have been one of his medicine songs. I believe he sang the song to give it away. He did not want to become 'too strong.' He was prepared to follow his dreams toward Yagatunne, the Trail to Heaven. I did not know, then, that a person could sing his medicine song only when death was near to him or to the listener. I did not know that the song had power to restore life or to take it away. I did not know he was giving up power the foxes gave to him in a time out of time, alone in the bush in

the 1890s. The song fell away and Japasa resumed his narrative. Johnny continued to whisper a translation:

My dad said he stayed out in the bush for twenty days.
Ever since that time, foxes have been his friends. Anytime he
wanted to, he could set a trap and get foxes.[13]

Ridington records two other stories told by Japasa: 'Indian
people from far and wide used to gather in the prairie country near the Peace River to dry saskatoon berries. They came
down the rivers in canoes full of drymeat, bear tallow, and
berries. They sang and danced and played the hand game in
which teams of men bet against one another in guessing
which hand conceals a small stone or bone. The other story
was about frogs who play-gamble, just like people. He said
he knew frogs because he once lived with them on the bottom of a lake.'[14]

What is central about this approach to time is its creation
of links between the reality of the everyday and the spiritual
reality that we might prefer to describe as mythic; the links
are the product of a person's power or knowledge. As Riel
and Tanner suggested in their discussions of foreknowledge
in hunting, so in this approach to the dimensions of time and
history the individual is enabled to live simultaneously in
what we perceive as two dimensions, whether labelled 'past
and present' or 'this world and the next.' According to
Calvin Martin: 'The key here is one's cultural conception of
time. Archaic societies maintain their strikingly vivid relationship with Nature because of their a priori commitment to
living in mythic, rather than in historic, time.'[15] This prior
commitment distinguishes traditional from contemporary
Canadian experience. It can be described as knowledge or
understanding of these other dimensions. Indeed, Ridington
concludes: 'For northern hunting people, knowledge and
power are one. To be in possession of knowledge is more
important than to be in possession of an artifact.'[16] And time
thus could be bent to fit the shapes established by spiritual
understanding.

The perception of space, particularly as nature and land, is as important as the perception of time in distinguishing between the traditional Aboriginal universe and today's cultural assumptions. The evidence begins in Aboriginal myths because, as is evident in the setting for the story told by Alestine's grandmother, telling stories and making speeches are among the most highly valued arts in traditional cultures. The land provides many of the characters: river and mountain, caribou and mountain goat, summer sun and winter snow offer context and actors for these tales. Whitefish and grouse and rabbits are the faithful supports of human existence. Other creatures take on the characteristics of humans. Thus for certain Aboriginal groups Bear is the warrior, Wolf the isolated spirit, Wolverine the outcast. In offering models of leadership, Crane is wise and Loon is ambitious. For Aboriginal storytellers, music might be evoked by the song of a bird, the winds of the forest, and the waves on a lake – sounds that are actually providing a backdrop as they speak. The rhythm of life is reflected in the succession of the seasons: in an Ojibwa story, harsh Winter wages war against gentle Summer, but neither secures permanent victory. These are not only the expressions of widely observed human experiences, although they do have that status, but they carry too the messages of specific locations on the earth's surface. James Bay and Hudson Bay and Lake Superior and Lake Winnipeg – their names today – figure in these stories; the Rocky Mountains and the Atlantic Ocean are boundaries of experience; and the distinctive combinations of nature's family in this part of the globe – mosquito and seal and magpie and muskrat – are unmistakable signposts of place.[17]

The rituals of hunting bands are drawn from hunting itself and are associated with special moments or stages in life, forecasts of game, and control of the weather. A typical rite of passage is the Mistassini Cree 'walking out' ceremony, which occurs soon after children learn to walk. It formally marks the moment when a child can begin to be active outside the dwelling. Young children are led straight out of the doorway of the household and along a path to a small decorated tree, where they perform a symbolic act such as shooting an animal or gathering fuel. They then circle the tree and return to the dwelling and family for a special meal. Each element in this ritual, including the walk itself, the tree (a vehicle of communication between human and animal), the meal, and the symbolic task, carries a spiritual message translated into the language of the land.

Cree dwellings often face a body of water and the rising sun, partly for physical reasons (prevailing winds, for example) and partly to permit communication with certain spirits of the natural world. The consumption and disposal of animals after a successful hunt also involve rules and rituals that convey the band's expectation of comparable harvests in the future, expressed in the languages of the hunt and the land. They imply a sense of belonging to the universe that is predictable and reassuring, but, in their use of the identifiably local characteristics of the boreal forest of northern North America, they denote also a specific sense of place.

Aboriginal maps or abstract depictions of the environment have rarely survived the passage of centuries. However, the few pieces of evidence that have been passed on reveal clues about the way in which different cultures shape the environment according to different principles. The canoe routes through northern Canada's intricate networks of rivers and lakes that Aboriginal people mapped at the request of fur traders were customarily measured in time rather than in spatial units, so that an easy downstream paddle of fifty miles and a difficult portage of eight miles might occupy the same space on a diagram because they required equal time to traverse. Such diagrams frequently ignored the distinction between river, portage, and trail, presumably because they were 'parts of a single communication network.' Moreover, and for similar reasons, small strategic creeks might be represented as equal to major water courses.[18]

It was a short step from depictions of canoe routes to depictions of hunting resources and, most essential of all, to depictions of the link between earthly spaces and the super-natural. The experience of Hugh Brody in northern British Columbia offers just such an insight into the hunters' spatial imagination. Brody was investigating Indian land use. He learned that members of the band might fish a particular pool or pick berries in a certain patch only once in five or ten years, but they remembered the location of the resource clearly and regarded it as theirs; the maps that they drew of their patterns of land use indicated clear boundaries between the customary hunting areas of one community and those of another.

The relation between these secular drawings of patterns of resource use and the spiritual depiction of the land was even more striking. The people with whom Brody lived, like the Indians in fur trade journals, described how the most skilled and perceptive hunters could dream the hunting trails, find the quarry and dream-kill it, then awake and collect the animal just as had been foretold. These men and women knew where the animals came from, where the berries grew, where the trails converged, and even where they originated; the best men and women – the strong dreamers – could map the animal trails and the berry patches and follow the trails to the points of convergence and from there, 'where the trails to animals all meet,' draw the route to heaven.

At a public hearing concerning plans for an oil pipeline in this area, after the speeches and the testimony were over and a community supper had begun, Brody witnessed an Aboriginal map reading that demonstrated the interconnection between the present world and another one:

> Jimmy Wolf's brother Aggan and Aggan's wife Annie brought a moosehide bundle into the hall. Neither Aggan nor Annie had spoken earlier in the day, but they went directly to the table at which the elders had sat. There they untied the bundle's thongs and began very carefully to pull back the cover.
>
> At first sight the contents seemed to be a thick layer of hide, pressed tightly together. With great care, Aggan took this hide

from its cover and began to open the layers. It was a magnifi-
cent dream map.

The dream map was as large as the table top, and had been
folded tightly for many years. It was covered with thousands
of short, firm, and variously coloured markings ... Abe Fellow
and Aggan Wolf explained. Up here is heaven; this is the trail
that must be followed; here is a wrong direction; this is where
it would be worst of all to go; and over there are all the animals. They explained that all of this had been discovered in
dreams.

Aggan also said that it was wrong to unpack a dream map
except for very special reasons. But the Indians' needs had to
be recognized; the hearing was important. Everyone must
look at the map now ... They should realize, however, that
intricate routes and meanings of a dream map are not easy to
follow. There was not time to explain them all ...

A corner of the map was missing and one of the officials
asked how it had come to be damaged. Aggan answered:
someone had died who would not easily find his way to
heaven, so the owner of the map had cut a piece of it and buried it with the body. With the aid of even a fragment, said
Aggan, the dead man would probably find the correct trail,
and when the owner of the map died, it would all be buried
with him. His dreams of the trail to heaven would then serve
him well.

This moosehide dream map was an extraordinary document
and artefact; it was at once a diagram of the region, a repre-
sentation of animal trails and other resources, and a sketch of
the path to eternity.[19]

What were the salient characteristics of the Aboriginal cultural order? The culture, first, was constructed on a particular
sense or dimension of time. Second, it postulated a unity of
experience among humans, the animal world, and the rest of
the natural universe – that is, it recognized an accessible,
direct link between this world and a dream trail. Third, it
relied on the land. These three points, which emphasize the

role of the land in traditional culture and the unity of experience between this world and another, constitute fundamental characteristics of a perception of time and space that differ from the conventional perceptions today. Fourth, human interactions within these distinctive time–space dimensions occurred in known locations – *this* river, *that* chain of mountains, *those* hunting sites – in a specific part of the world. The unmistakable link between people and place dating from time immemorial, when combined with their cultural distinctiveness, substantiates Harold Innis's sweeping judgment about the Aboriginal foundation of Canadian institutions: the very community sites, the places where people now live, have been occupied from time immemorial – that is, from the earliest versions of time and space – by the ancestors of individuals who live in these locations today.

III

Robert Bear, then in his eighties and an elder on the Little Pine Reserve, offered a succinct summary of world cultural history for the museum in Prince Albert, Saskatchewan, when a new exhibit was being prepared in the 1980s: 'The Indian was the first here and this is where he was created; the white man was created overseas and came here afterwards. The Bible we were given was nature itself; the white man was given a book.'[20] In other words, as the mythic narratives and the dream map reveal, many of the crucial elements in hunting band life in the Americas were conveyed from one generation to the next by oral tradition. Dreams and diagrams were not recorded for posterity in the pages of a book but rather depended on the insights of new dreamers to revitalize band beliefs and on the words of wise elders to aid in the interpretation of complex visions. Aboriginal communication before European contact relied on the spoken word, and to a considerable degree the oral message has remained the vehicle of religious perspectives.

Can one seriously contend that the culture sustained by

oral traditions has survived in our literate age? Can it actually constitute a foundation for Canadian institutions? Knowledge in a society that relies exclusively on oral transmission of its culture is extremely fragile, writes James Axtell, 'because its very existence depends on the memories of mortal people, often specialists who are entrusted with major portions of the corporate wisdom. Abnormally high death rates in one or more generations sever the links of knowledge that bind the culture together.'[21] Moreover, for four centuries, Aboriginal spiritual or cultural beliefs have been the targets of European criticism and indeed of direct assaults by European missionaries and teachers. If the Aboriginal cultures drew their strength from the land, they were also undermined by the very characteristics – the orality and the land-based metaphors – that sustained them.

The Aboriginal peoples' reception of print in the seventeenth century shows what a devastating impact the Europeans' literacy had on their society. One story must suffice. The Recollect priest Gabriel Sagard reported that, when he was on a trip to Quebec with some Hurons and discovered that one of the canoes was leaking, he sent a note to a priest colleague back in the village whence they had come to ask for another canoe: 'When our canoe arrived,' Sagard wrote, 'I cannot express the admiration displayed by the natives for the little note I had sent to Father Nicolas. They said that the little paper had spoken to my brother and had told him all the words I had uttered to them here, and that we were greater than all mankind. They told the story to all, and all were filled with astonishment and admiration at this mystery.'

To the Aboriginal observer, as James Axtell says, print 'duplicated a spiritual feat that only the greatest shamans could perform, namely, that of reading the mind of a person at a distance and thereby, in an oral context, foretelling the future.' This was a feat identical in form, in the Aboriginal cultural perspective, to that of the greatest hunters who could foretell the kill of an animal or map the trails to

Heaven.[22] It was identical too to the story about the 1995 rescue of the roadgrader operator.

The First Nations were introduced to the powers of print by missionaries, particularly by the Bible, the missionaries' means of reference to their deity. The existence of such a holy work redoubled the Europeans' authority, in the view of some Aboriginal people, because it seemed to promise that the priests had continuous contact with their God. These Aboriginal 'radicals,' as they might be labelled to distinguish them from 'conservatives' who wished to retain inherited or traditional perspectives, argued that the priests possessed tangible proof of their doctrines: 'Christian doctrine was immutable,' they could claim, 'and therefore superior to native religious traditions, because it was preserved in a printed book as it had been delivered by God.' Axtell suggests also that print was a crucial weapon in the missions of New France: 'Wherever they went, the Jesuits hammered home the point that "the Scripture does not vary like the oral word of man, who is almost by nature false." Huron converts had been so persuaded by their resident Black Robes that they would interrupt the traditional history recitations performed at council elections to make the same argument ... When Father Joseph Le Mercier reported to his superiors in 1638, he drew up a list of what inclined the Hurons to Christianity. At the top of the list was "the art of inscribing upon paper matters that are beyond sight."'[23]

Literacy and Christianity constituted a powerful challenge to Aboriginal cultures and were wielded like weapons within them. Still, it would be wrong to think that the cultural link between past and present in Aboriginal society has been broken by the twin forces of literacy and Christianity. A convincing display of resistance is the Andre family itself: grandmother, the story she tells, and the listeners who absorb her message exemplify cultural continuity.

It is obvious that orality, as a dominant mode of communication, has been vanquished by literacy and print. One cannot recover a world without clocks, books, lists, and codified laws and regulations. But Aboriginal culture has not been obliterated. Aboriginal people are not simply like other Canadians. Why not? Their belief in an unbroken chain between past and present remains unshaken, as is evidenced by a thousand stories about Aboriginal cultural resurgence in contemporary Canada.

The challenge is to understand how this continuity has been maintained. If the explanation relies on a culture built on orality and traditional time–space dimensions, how are these insights communicated today? I suggest that the answer lies in politics. The community's culture finds its expression in the political determination to survive, to restore lost resources, and to build a better world for children such as little Lisa in the Andre family's fishing camp. Cultural continuity is ensured by the effective communication of political causes.

IV

This final stage of my discussion of Aboriginal culture concerns political activity. It proposes that cultural continuity has been maintained by the consistency and tenacity – the sheer stubborn immovability – of Aboriginal people over the past several centuries. Surely their efforts reveal evidence of effective communication from one generation to the next.[24] They offer evidence too that Aboriginals' confidence today rests in their awareness that their greatest cultural resource lies in their peoples' unbroken habitation of this territory from time immemorial to the present. As Marshall Sahlins observes about the worldwide evolution of Aboriginal cultures, *'the strongest continuity may consist in the logic of the cultural change.'*[25]

It is not easy to penetrate the actual workings of the Aboriginal political community. However, Hugh Brody's research in a British Columbia hunting community during the 1970s has provided some remarkable insights into Aboriginal communications. His experience tells us that Aboriginal relations with the natural world – marked as they are by a religious dimension – are built on a distinctive cultural foundation. His writing also shows why European Canadians have had such difficulty understanding the very different process of consensus-building among Aboriginal people.

When he first joined the hunting community near Fort St John, Brody was invited to accompany Joseph Patsah and several other men on a hunting and fishing trip. The process of planning and decision making for this expedition, so 'muted' as to be perplexing and even irritating to a white participant, seemed to ignore the threat of hunger and the need for precision and thus to embody all the carelessness that some observers might perceive in Native lives. Rather than weighing factors, calculating costs, or advocating a course of action,

> Joseph and his family float possibilities. 'Maybe we should go to Copper Creek. Bet you lots of moose up there.' Or, 'Could be caribou right now near Black Flats.' Or, 'I bet you no deer this time down on the Reserve ... ' Somehow a general area is selected from a gossamer of possibilities, and from an accumulation of remarks comes something rather like a consensus. No, that is not really it: rather, a sort of prediction, a combined sense of where we *might* go 'tomorrow.' Yet the hunt will not have been planned, nor any preparations started, and apparently no one is committed to going. Moreover, the floating conversation will have alighted on several irreconcilable possibilities, or have given rise to quasi-predictions. It is as if the predictions are about other people – or are not quite serious ... come morning, nothing is ready. No one has made any practical, formal plans. As often as not – indeed, more often than not – something quite new has drifted into conversations ... The way to understand this kind of decision making as also to live by and even share it, is to recognize that some of the most important variables are subtle, elusive, and extremely hard or impossible to assess with finality. The Athapaskan hunter will move in a direction and

at a time that are determined by a sense of weather (to indicate a variable that is easily grasped if all too easily oversim-
plified by the one word) and by a sense of rightness. He will
also have ideas about animal movement, his own and others'
patterns of land use ... But already the nature of the hunter's
decision making is being misrepresented by this kind of list-
ing. To disconnect the variables, to compartmentalize the
thinking, is to fail to acknowledge its sophistication and completeness. He considers variables as a composite, in parallel,
and with the help of a blending of the metaphysical and the
obviously pragmatic. To make a good, wise, sensible hunting
choice is to accept the interconnection of all possible factors,
and avoids the mistake of seeking rationally to focus on any
one consideration that is held as primary. What is more, the
decision is taken in the doing: there is no step or pause
between theory and practice. As a consequence, the decision
– like the action from which it is inseparable – is always alter-
able (and therefore may not properly even be termed a decision). The hunter moves in a chosen direction; but, highly
sensitive to so many shifting considerations, he is always
ready to change his directions.[26]

Brody's insight into the relation between the thought of
an individual and the action of a group underlines the complexity of the Aboriginal world-view. It suggests the contingency of knowledge, according to the Aboriginal estimate,
in relations between humans and the natural world. And it
demonstrates the remarkable inclusiveness with which
Aboriginal people approached collective action. Brody's perception is in harmony with Adrian Tanner's account of divination among Mistassini Cree hunters and Louis Riel's
discussion of foretelling. All three narratives reinforce the
image of a continuing oral culture. They emphasize community-wide discussion, the necessity of collective readiness for
action, the absence of coercion, and the evaluation of intangible as well as tangible factors that might shape events.
They make thought, action, and relations with the environ-

ment inseparable. Grandmother Andre's decision to take part in the film-making and to don her remarkable coat, after a period in which she was uncertain, illustrates this contingent approach.

Decision making in Aboriginal cultures took into account all human and environmental influences. In such discussions, culture and land were inextricably bound together. The two must be in harmony if wise decisions are to result. However, such a method of making decisions does not fit easily into modern Canada, with its vast territory, sizeable population, and more insistent, compartmentalized links between people.

Canada's Indian Act, passed in 1876 and amended often in the following decades – and still in place in the 1990s as the key legislation defining government-Aboriginal relations – reveals the failure of Canadian institutions to comprehend Aboriginal circumstances. The act imposed a system of government on the Aboriginal peoples that is in retrospect extraordinarily paternalist and racist. It is also, in theory and practice, patriarchal. It constitutes a direct attack by the government on Aboriginal political arrangements. At one time, it permitted the federal government to control the selection of Indian community leaders, it conferred control of community membership on that government, and it introduced European-Canadian 'Indian agents' (really community supervisors), who could pass judgment on almost every aspect of band decision making. Within a decade of its passage, Ottawa was using the Indian Act and supplementary legislation to prevent Aboriginal people from expressing their beliefs, pursuing their traditional economy, and asserting their political rights.[27] Far from merely 'civilizing' their material conditions or outward appearance – the declared goal of previous legislation – the new laws aimed to bring about Aboriginal assimilation. As the chief federal civil servant in the Department of Indian Affairs wrote in 1920, 'Our object is to continue until there is not a single Indian in Canada that has not been absorbed into the body politic, and there is no Indian question, and no Indian Department.'[28]

The Aboriginal peoples' response has consistently focused on land policy.[29] Big Bear and several other prairie Indian leaders, for example, demanded significant revisions in Treaties Four and Six (signed in what is now Saskatchewan in 1874 and 1876). Their goal, aside from the obvious desire that Ottawa actually fulfil the terms of these treaties, was the establishment of large contiguous reserves in place of the small plots scattered across the North-West. Why? The Indian leaders desired the negotiating power, one might suppose, that comes with the concentration of so many potential soldiers in one area. But, in this case, Big Bear and Little Pine were worried as well by the disappearance of the great buffalo herds; perhaps they wished to ensure that sufficient tracts were preserved to permit the survival of hunting economies during the transition to another way of life. Land, the foundation of the hunting band, was their concern.[30]

This same preoccupation with the relation between land and culture was evident in the thinking of Louis Riel. Riel had clear principles and a persuasive strategy for Métis. He estimated the value of western lands in Native possession at 25 cents per acre for the Métis. He suggested that land in the possession of Indians should be valued at a lower rate – half to two-thirds that for the Métis, because Indians exploited its resources less intensively. Riel proposed creation of a development fund to compensate for the transfer of land from Aboriginal to Canadian-government ownership. The interest paid on the fund would, he suggested, cover the current needs of the First Nations, pay for their educational and health institutions, and support their adaptation to European-Canadian ways. Riel, a Métis, was not especially concerned to preserve the traditional culture; but he would use that society's crucial resource – land – to sustain the next generations of Aboriginal Canadians.[31]

Land has been the paramount issue in relations between government and First Nations for the past century. The files

of the Indian Affairs and Interior departments are stuffed
with petitions and debates concerning particular local claims
and surprisingly, given the government's wish to extinguish
Aboriginal rights, about vast 'comprehensive claims' to land
and resources. Without their own land, Aboriginal people
lack an essential element in life. As a university anthropologist explaining the crucial role of reserves in Aboriginal culture told a parliamentary committee in 1947, 'The Indian
belonged to the land, land did not belong to the Indian.'[32] A
contemporary Ojibwa writer notes that her grandparents
thought in similar terms: 'Even after 40 years, I can still hear
their voices. "Remember your connection to the land. It's the
land that makes us look, act and talk the way we do," my
grandfather says in Anishnabe.'[33]

What view of land is represented here? Are we talking
about resource income and real estate speculation alone, or is
something more profound at work? A newsmagazine article
on Matthew Coon Come, the leader of the Grand Council of
the Cree of Quebec, offers an insight. When he was an under-
graduate at McGill University, Coon Come asked his father
to teach him about the land: '"I was a typical college kid," he
says, recalling how he had arrived in the bush armed with a
detailed topographical map of the area he and his father,
Alfred, were about to explore. "The first thing my Dad did
was tear that map into tiny little pieces," Coon Come continues. "He said I was committing the white man's mistake,
making plans for the land without ever setting foot on it,
without ever getting a feel for it."'[34]

Canadian Aboriginal people have continued to assert their
view of the country's political arrangements by speaking of
the relations among culture, religion, land, and self-government. In retrospect, the consistency and firmness of Aboriginal resistance to Ottawa's paternalism must be humbling to
outsiders. Aboriginal leaders have endured jail sentences
imposed for offences ranging from 'treason-felony' to conducting dances of worship and celebration. Throughout the
century after 1876, they demanded the right to make their own

decisions concerning community and individual life. Aboriginal presentations to a parliamentary inquiry in the late 1940s expressed this priority clearly. Responses to the Trudeau government's 'White Paper' of 1969 reiterated these claims.

The Aboriginal campaign for constitutional recognition reached another frustrating impasse in the 1980s, expressed with dramatic force at a pivotal moment in the country's constitutional history. In 1990, Elijah Harper, an Aboriginal NDP member of the Manitoba legislature, working with representatives of all of Canada's Aboriginal people, prevented the passage of a complex constitutional package, the Meech Lake Accord. Almost simultaneously, a conflict between an Aboriginal group and a Quebec municipality over a tract of land near Oka escalated into an armed standoff in which a Quebec police officer was killed. The Canadian Aboriginal community was galvanized into action. The chief of the Assembly of First Nations, Georges Erasmus, in defending Harper and the Indians of Oka, spoke compellingly about the Aboriginal *place* in Canada. In a moment of high drama, when speaking to a protest rally that included hundreds of Aboriginal people, he referred implicitly to the relations between place and immigrants to Canada by declaring: '*This is our land. We have no other land.*'

The problems were addressed again in the Charlottetown Accord, another complex constitutional reform package that, among many other things, would have set up a version of Aboriginal self-government. This accord was submitted to a national referendum in 1992 and was rejected. Aboriginal people were divided in their view of the measure. In 1996, a Royal Commission on Aboriginal Peoples concluded several years of work with the release of a four-thousand-page report containing history, analysis, and recommendations. In 1997, the federal government issued a formal statement of reconciliation as a step in the reconstruction of relations with Aboriginal people. The struggle continues.

How are Aboriginal people fundamental to the growth of

Canadian institutions, to return to Harold Innis's thesis? On the surface, the answer must lie in the First Nations' association with these northern lands. As Innis himself demonstrated, the territory of beaver populations, the interlocking pattern of the northern river systems, and the longtime Canadian reliance on the export of this fur staple to Europe through distinct corporate channels ensured that the destiny of the northern half of North America would be separate from that of the southern half.[35] As many observers have also noted, the relatively greater harmony of the Indian-European exchange in this territory and the emergence of mediating communities – the Métis – also were by-products of the European adaptation to Aboriginal culture. In many parts of northern North America, especially in the west and the north, Europeans accepted and worked within Aboriginal cultural conventions. In this way, Aboriginal people guided the creation of a separate country.

This is, however, a European-Canadian explanation. A more comprehensive and Aboriginal view relies on the elements of First Nation culture. In their sense of time and space, in their ritual and myth, and even in their decision making, Aboriginal people used the language of their environment. Land – this particular tract that we recognize as northern North America – was the shaping force in Aboriginal existence. Because they developed their distinctive cultural dimensions of time and space in relation to this place, their adaptations must constitute their founding role in Canadian experience. Because today's Aboriginal people represent an unbroken chain of human occupancy from time immemorial, their sense of this continuity must be counted as a foundation of Canadian identity. Because they have shared their inheritance with the newcomers, they are, as Innis wrote, 'fundamental to the growth of Canadian institutions.'

PART TWO

Textual-Settler Societies

3

Elizabeth Goudie and
Canadian Historical Writing

Newcomers from other lands reconstructed the existing society of northern North America. Aboriginal people did not disappear – far from it – but the community, including the First Nations people who were adapting to it, diverged from its former trajectory.

Despite obvious differences, including language, religion, and national origin, the many groups of new settlers had much in common with the various groups of First Nations. Their daily life, for example, was constructed on 'traditional' dimensions of time and space. However, theirs was a 'textual' society, in which written texts supplemented oral communication, even though individual families might have little access to the fruits of this civilization. Thus, although they lived daily with the obvious truth that the rhythms of nature shaped their lives, they also acknowledged the power of new modes of communication, especially print. They dealt every day with the fact that some of their work contributed to international trading enterprises over which they had no control and about which they knew little. Finally, whether trappers, woodcutters, fishers, farmers, hunters, or more likely combining these activities, they were also children and parents, women and men, and in such family roles they shared similar experiences because the household was the crucial unit of production and survival in every region.

Elizabeth Goudie at age sixteen, taken at Sebaskachu, Labrador, 1918
(above left), and at age sixty-eight (above right).

This chapter examines conventional interpretations of 'set-tler' history. The four approaches – genealogy, survival, fron-tier, and staple – have become embedded in Canadian popular culture and do offer sensible views of the pioneer past. However, this chapter argues that none, however wor-thy it may once have been, offers a convincing story today. And it outlines an alternative. Chapter 4 shifts from geneal-ogy and economy to culture, like chapter 2, and explains the connections between present and past with reference to peo-ple's experience of time, space, and political action. Once again, as in part 1, I am interested in discovering how these diverse peoples also were fundamental to the growth of Canadian institutions.

I

Let us consider the life of Elizabeth Goudie, who was born in 1902 in Labrador and trapped, gardened, and fished in iso-lated circumstances near the Labrador coast.[1] After her hus-band died, Elizabeth, then in her sixties, took up pencil and scribblers to write the story of her life for her grandchildren

– a manuscript that provides a touchstone for this chapter and the next. She had grown up in a village on the shore of Hamilton Inlet, one of five villages of three to five families in size, and had had the privilege, as she wrote, of going to school for several brief sessions, which enabled her to acquire perhaps a fourth-grade standing. Her family subsisted on fresh meat and fish, peas and beans, rolled oats porridge and tea sweetened with molasses. On special occasions, such as Christmas and Easter, the molasses gave way to a rare treat, milk: 'I tell you we were glad to see Christmas or Easter.' In spring, for a full month, her father prepared a drink made from boiled spruce, juniper, molasses, yeast, raisins, and rice 'to clean up our blood.'

Elizabeth commented in her memoir: 'There was not much around except lots of hard work.' As a child, she looked after her younger brothers and sisters, prepared meals, made and mended clothes, manufactured soap, and did the hundred other tasks necessary to sustain the family. Of parental discipline, she wrote: 'Perhaps they went to the extreme a little, yet I think a lot of their rules helped us along the way. Our parents were always reminding us to be honest and truthful and kind to others ... these were some of the things that helped us to be contented with one another and with what we had.' Because the family was very poor, Elizabeth had to accept wage work away from home at the age of fourteen ($2 a month plus bed and board, later raised to $4 a month plus used clothing): 'You had to do everything by hand: scrub, wash, bring wood and water, help to cook and mend clothing. We never had a dull moment.'

Elizabeth met her future husband when she was fifteen. They became friends and 'went out together when I had my afternoon off. We enjoyed the summer very much.' Jim visited her briefly the following year, just before he went to St John's to train in the army, and he returned the following winter, when, having been too late to join the Newfoundland troops before the signing of the Armistice on 11 November 1918, he had caught the mail steamer heading up the coast.

He then travelled overland by foot and dog team, arriving at his home village of North West River in February 1919. His was the first report in the district of the war's end.

Elizabeth and Jim visited together three times in the next year. They decided to become engaged in February 1920, just before Jim headed back to his trapping district. Having survived a dangerous plunge into icy waters in weather of minus thirty degrees, Jim walked forty miles on snowshoes to sell his furs and another twenty-five miles to secure the consent of Elizabeth's parents to their marriage. He then sent a dog team to inform his prospective bride. Elizabeth noted that she was then eighteen (Jim was twenty-seven): 'I had to make some of my wedding clothes, but I bought my dress from Mrs Paddon. It cost me five dollars. It was a white dress with a little blue strip in it. It is forty-three years now since I was married and I still have my dress.'[2]

The newlyweds moved north to the Hopedale district in 1922 in search of hunting and trapping grounds on which to build their new life together. Elizabeth learned to shoot and fish under the tutelage of her husband and her future sister-in-law. Jim cut trees to build their cabin and sawed the wood for household fires. Elizabeth made clothing for him – two sets, including mitts, socks, and boots: 'We always prepared extra things in case of accidents like fires or losing the canoe in the river.' Elizabeth commented about their labour: 'Every man and woman in Labrador in those days who was able to work at all had to be a jack-of-all-trades. There was no other way out.'

A month's labour was necessary to outfit her husband for his trapping expedition: 'We didn't have much time for fun. We worked early and late. The day came for him to leave. That was a very sad day for a trapper's wife.' Elizabeth always hated this moment of parting. 'The most dreadful thing' about her lot, she wrote, was that she heard nothing from him for up to three months: 'The first year I was married was a very lonely one.' However, her reflections on this period, written when she was past her sixtieth birthday, con-

Henry Blake, Donald Baikie, and Jim Goudie, North West River,
Labrador, 1918.

tained more positive than negative judgments: 'But there was something about that life that is hard to put into words. It was a life not full of people or what people could offer you. You would rise in the morning. There were no people around you, but every day you had something to make you happy. We were satisfied ... We always said if God wants us to live he will provide for us, as long as we do our part. We were content with that thought.'

During their second year of marriage, Elizabeth became pregnant. Rather than stay alone while Jim was away trapping, she travelled to her mother's home, where a midwife aided in the delivery of her first son: 'He weighed eight-and-a-half pounds and was just like his daddy. He had a dark complexion and I was proud of him.' Horace was the first of nine children born over the next twenty-five years, seven of whom survived to adulthood.

Illness punctuates Elizabeth's memoir as sharply as it must have daily life. Bruce, her fifth child, was burned badly about the face and upper body in a cabin fire, but she nursed him back to life only to lose him to an unnamed disease just six months later. Her husband nearly died from an axe wound sustained on a trap line far from settlements and medical care. Her seventh child contracted whooping cough and died only three weeks after birth.

The Second World War and air transport changed the life of the Goudie family. Elizabeth begins this section of her story with the declaration: '1939 brought the old life of Labrador to a close.' Air travel, defence bases, and radio became part of the family's experience. Her response to some of these 'firsts' was evident in her note on events in 1943, when her family, transported by dog team, left Mud Lake for Goose Bay. A truck was waiting to take the locals on an adventure: 'When I first saw the truck I didn't feel much like getting on board, but because some of my children were afraid, I had to be brave. Everyone got on and we started towards the air base. The road was rough but the driver took extra care.' They were taken into one of the new buildings, where the airmen gave chocolate and gum to the children and then 'turned off the lights and turned on the show. When it came on we couldn't believe our own eyes and ears, hearing people talking and seeing them moving on the screen. I thought a lot about it all after I got home that night.'

The Goudie family accepted a more sedentary life after the war. Jim secured work at the military base, Elizabeth worked in their new house, and their four younger boys stayed in school. Jim died in 1963, at the age of seventy, and Elizabeth commenced her memoir later in the year. The scholar who helped her with the editing of the manuscript described her as belonging to 'the last generation of real pioneers.'[3] This judgment underlines the great span of time – from the early seventeenth to the mid-twentieth century – during which the new settlers' cultural dimensions prevailed in parts of northern North America.

The idea that Elizabeth Goudie's life and memoir represent important influences in contemporary Canada is not far-fetched. True, her story is a chronicle of hardship and endless labour endured in an isolation unlike the experience of most citizens today. Still, it should be possible for every Canadian to claim Elizabeth Goudie as a founding citizen of their community and as a force in shaping their own identities. Given that this will seem an unusual contention, especially to young people raised on video games and the internet in Canada's metropolitan centres, how can the necessary cross-cultural communication be established?

In the generation before ours, a generation whose intellectual horizon was sketched between the 1920s and the 1960s, the writing of Canadian history relied on four chief means of establishing contact between present and past and of asserting a logic and continuity to the Canadian nation. The first of these, genealogical, assumed that each generation descended literally – as children and grandchildren – from the founding fathers and mothers. Those who were moved by the environment itself, including Northrop Frye and the young Margaret Atwood, suggested a second – that sheer survival was the pioneers' greatest achievement. Those who read the works of a young Arthur Lower or S.D. Clark discovered a third model, wherein the frontier was Canada's defining experience. The fourth, drawing on the writings of Harold Innis and Donald Creighton, declared that the export of staples – fish, fur, timber, wheat – provided the foundation of Canada. Each of the history teachers and writers of that generation – MacNutt, Ormsby, Thomas, Cooper, Tallman, Morton, and the rest – found sufficient authority in these explanations to sustain their writings and teachings for a generation of students and readers.

These approaches seem less than convincing today for two reasons: the new social and regional history has altered our knowledge of the past; and cultural changes in recent decades have reconstructed the Canadian audience. None the less, the popular conviction that there is continuity between the pioneers of yesterday and the citizens of today is a vital element in contemporary life. Our challenge is to understand this continuity and thus to explain why settlers too are central to the construction of a Canadian historical narrative. Can one establish a common history spanning the experience of Aboriginal peoples and settlers? I believe that Elizabeth Goudie and the thousands of settlers like her show why the answer is 'yes.'

It is true that many Canadians descend from pioneer settlers. An exclusively genealogical interpretation of the settlers' place in contemporary Canadian life, however, would be no more satisfactory than a view of the Aboriginal role in Canadian history that emphasized only the family links between past and present generations. As with the Aboriginal experience, so in this chapter, I argue that genealogy is a necessary but not a sufficient explanation of the power of first settlers in shaping today's Canada.[4]

What is wrong with the genealogical approach? I suggest that it has been undermined by a new version of social history that introduces Aboriginal people into the story and demonstrates that the Métis played a profound role as intermediaries between European and Aboriginal people. Moreover, genealogy does nothing to address the hundreds of thousands of Canadians who have arrived in the country since 1900, especially those whose families originated outside Europe. An inclusive historical synthesis will have to find authoritative personal reasons that link these people to early Canada.

The immigrant settlers who arrived in northern North America were not as different from their Aboriginal neighbours as is sometimes supposed. Moreover, they relied on the original inhabitants to a degree that is often underestimated. Elizabeth described herself in her memoir as the daughter of a trapper. However, as she explains very carefully in the first chapter, 'Family History,' her great-greatgrandmother was an Inuit woman, an orphan, who married Ambrose Brooks, a fisher of English origin, in North West River, Labrador, in 1806. Several children from this partnership found jobs in the local operations of fur trade companies. They in turn intermarried with Scotch, FrenchCanadian, English and Aboriginal people. Thus, in recording her place in the family story, Elizabeth gives us a valuable insight into the history of her community: 'So the Hamilton Inlet was populated by the white people who came out here from Scotland, England and Canada, so we were told. They married among the Eskimo and Indian people. According to my parents this happened about two hundred years ago.' The society to which Elizabeth belongs is, in other words, partly European and partly Aboriginal.

The mixing of heritages also appears in discussions of economy and technology among these first settlers of Labrador. Thus Inuit household technology and Montagnais or Naskapi trapping and hunting experience merged with European practices. A visitor described the settlers as 'a unique race with oddly combined cultures; Scotch Presbyterian by religion, old English in speech and many customs,

Eskimo when it comes to seal fishing and dog driving, Indian in their ways of hunting and their skill with canoes ascending the big rivers bound for the trapping grounds far in the country.'[5]

What are the characteristics of that society which permit us to draw a line, however permeable, between traditional Aboriginal society and later versions of Canadian society? Elizabeth Goudie herself sketched certain boundaries. She implied that they, the Aboriginal people, differed in language, in religion, and in the society – the company – they kept. Some individuals, including her ancestor, the Inuit woman who married Ambrose Brooks, moved from one culture to another without apparent difficulty, although Brooks felt compelled – or so it is recorded in the family history – to teach the Inuit orphan 'enough English to say the Lord's Prayer' before he married her. The moments of education that took place across the cultural divide, as well as the acts of generosity extended by one community to another, were numberless. Certainly, Elizabeth gave and received in equal measure in her relations with the Aboriginal people who appear in her memoir. On one occasion, an Inuit household gave her entire family shelter. On another, she provided shelter and food at the risk of contracting disease as she assisted some Aboriginal travellers who were near despair.[6] Thus the points of separation were fewer, and the points of common experience more numerous, than today's mythology suggests.

The Métis, like Elizabeth, played a profound role in the history of northern North America. They were the intermediaries between European and Aboriginal people. The demographer Jacques Henripin has estimated that about 40 per cent of contemporary *pur laine* Québécois have some Aboriginal ancestry. A similar story can be told for Manitoba at the moment of its entry into Confederation in 1870, when over 80 per cent of the people in the Red River Settlement had partly Aboriginal roots. The presence of such new peoples, the Métis, from coast to coast in northern North America, speaking combinations of an Aboriginal and a European language – French or English or Gaelic or Russian – demonstrates that intermarriage and the intertwining of cultures lasted not just for decades but for several centuries. The development of merged communities influenced relations among individuals and families much more than is commonly credited.

This view has implications for today's Canadians. The genealogical school of heritage interpretation insists on all honour being paid to the founding fathers and mothers. Known to historians as 'filiopietism' (literally, praise or worship owed to the founders by dutiful sons and daughters), this stereotype of the past can be found in the literature, song, and dance of every cultural group. Such folk arts express the descendants' veneration of the group's first arrivals in northern North America. They also remove the pioneers from today's world by making their feats unrepeatable. They exaggerate the pioneers' qualities precisely because today's observers want to assert the exceptional quality of their own identity. Thus song and story declare that the founders endured hard times, sacrificed, and worked very hard as they adapted to a new physical and cultural setting. Inevitably, this filiopietism is a narrow loyalty, usually reserved for the 'pure' genealogical heirs within a community.

The idea that Canada's settlers – the founding fathers and mothers – descend from mixed-race peoples and from societies made possible by mixed-race peoples has not been part of filiopietist discussion. However, Europeans relied on Aboriginal people; the Métis occupied a pivotal role in history as intermediaries between the two. It is time that Canadians acknowledged the extensive contact between pioneer settlers and Aboriginal people. It is time to add up the intermarriages and the many occasions of cross-cultural borrowing and lending. If today's Canadians can respond to Elizabeth Goudie's experience by asserting that she, as a settler, belongs in a chain that extends to the present, then the construction of that bridge between Aboriginal and later arrivals is under way. Elizabeth represents an inclusive Canadian experience. Hers is not just a genealogy-based link between a specific descendant and a specific immigrant. Rather, viewed as a historical force, Elizabeth represents the amalgamation in a single narrative of Aboriginal and settler histories.[7]

What of the other three interpretations of Canadian history? The people's stereotypes of the pioneer past always include a dollop of dismay, as the second approach – survival – demonstrates. Northrop Frye's conclusion to *The Literary History of Canada* (1965) described the European traveller's entry into the St Lawrence as 'unforgettable and intimidating' and likened the experience to 'a tiny Jonah entering an inconceivably large whale.' According to Frye, the experience of Canadian first settlers was quite unlike that of Americans: 'To enter the United States is a matter of crossing an ocean; to enter Canada is a matter of being silently swallowed by an alien continent.' He claimed that Canada was 'full of wilderness ... the wilderness was all around one, a part and condition of one's whole imaginative being.' Students of literature have often quoted Frye's generalization: 'One wonders if any other national consciousness has had so large an amount of the unknown, the unrealized, the humanly undigested, so built into it.' In their mind's eye, Frye wrote, the first settlers inhabited 'small and isolated communities' wherein 'distinctively human values' were challenged by 'a physical or psychological "frontier"'; their cultural expressions exhibited a collective consciousness that Frye called a 'garrison mentality.'[8]

Although little has been made of the fact, the same theme resonates in the work of historians. In his conclusion to *The Canadian Identity*, W.L.Morton asserted that 'the central fact of Canadian history' must be the development of a nation in 'one of the largest, harshest, and most intimidating countries on earth.' Canadians had 'confirmed the essentials of the greatest of civilizations in the grimmest of environments.'[9] Morton claimed: 'The heartland ... of Canada [is] one of the earth's most ancient wildernesses and one of nature's grimmest challenges to man and all his works.'[10] The achievement of the first settlers, according to Morton, was that they had preserved the social values of the Old World despite difficult conditions in the New. Like Margaret Atwood, whose *Survival: A Thematic Guide to Canadian Literature* influenced a generation of students, Morton saw in the experience of the first settlers both the physical challenge and the narrow escape.[11]

The 'frontier' approach to the pioneer settlers – a third school – looks at the other side of the filiopietist coin. Rather than celebrating the settlers' heritage in previous homelands, the frontier hypothesis concentrates on the power of the North American environment to mould individuals and groups. The freedom of a relatively unformed community, it is said in this interpretation, permitted construction of a society in which ability and perseverance, rather than inherited class lines, book-learning, and old wealth, shaped government and economy. Today's western and northern populists often identify their cause with frontier images.

A fourth stereotype of Canada's pioneer past, one that has informed Canadian university history courses for over half a century, is often called the 'Laurentian thesis.' It locates a unifying thread of national experience in the 'staple economy' of northern North America. The first settlers may have harvested fish, fur, timber, and wheat in different regions and eras, according to this view, but their exports belonged to a single economic and cultural category – the staple. Such products needed little processing, could be shipped in bulk, travelled the same west-to-east routes of water transport (notably the St Lawrence, hence the 'Laurentian' title), and were sold in European markets under similar institutional arrangements. Over the course of several centuries, the communication systems constructed to convey such products divided the northern part of the continent into two social and political entities, Canada and the United States, along the lines set down by the river systems and natural resource zones. The staple export – its character as an economic good and its power as the founding purpose of a community – provided both a theory of economic growth and a satisfying explanation of Canada's separate national existence.[12]

Not only have these popular stereotypes of settler history endured, but they probably seem obvious to today's readers. However, if there is any truth in the platitude that every generation must write its history anew, then the stereotypes are due for revision. And that is indeed what historians have been doing during the past thirty years. Like the genealogical notion, the survival, frontier, and Laurentian stereotypes have been toppled from their perches. Now it is time for ordinary Canadians to catch up with historical research and to consider its implications.

II

What's wrong with the survival myth? A generation of social history research demonstrates that, although literary works may have emphasized the challenges of nature, experience like Elizabeth Goudie's was closer to the norm. Yes, many of these families lived with very few material possessions, and yes, on rare occasions, they came very close to starvation. Nevertheless, their existence was reasonably secure, and they perceived it in that light. They lived on the fruits of the land, fruits that were available to all, including ocean and freshwater fish, the animals and wood supplies of the boreal forests and prairies, and the berries and plants and migratory fowl of a rich and diverse environment.

A Métis woman recalled how, when she married in the 1860s in what is now Manitoba, she and her husband were producing nearly all of their material requirements. Her Red River home, like those in the eighteenth-century Métis communities in the lower Great Lakes, was constructed of logs cut locally and relied on a clay fireplace and chimney built on the spot. Her family's gardens and grain plots and hay fields stretched back from the river bank in the French Canadian and Scottish fashion. The buffalo hunt supplied them with quantities of meat, pemmican, and robes. Their clothing, also of domestic manufacture, combined European and Indian styles and included the shirt and vest of the former, the rawhide leggings and moccasins of the latter, and a colourful wide sash or belt, woven from wool (the *ceinture fléchée*), that was French Canadian in style. Like their Aboriginal parents and mentors, the Métis had adapted to the land in which they lived. This experience reflects exactly the story told by Elizabeth Goudie about her life in Labrador c. 1920–39. Such settlers not only made their homes and furniture from local resources, but they took pleasure in local materials and forms.

The households of New France, too, harvested adequate supplies of food and found sufficient resources to built shelters and to make clothing. The diet in the habitants' homes, rich in bread and vegetables, fish and game, was far superior to that of their French counterparts. Although mortality among infants (i.e., children under one year) was as high in New France as in Europe, once the dangers of the first year were surmounted, life expectancy was distinctly greater in the New World parishes than in the Old.[13] Most of the family's material goods – carts, furniture, even bedsheets and shoes – were made at home. Almost all the food was prepared from the products of farm, fishery, and hunt.

The image of Canadian wretches huddling against the arctic cold must be contrasted with European travellers' reports on habitant houses: 'In winter, by the aid of a stove, they [the houses] are rendered completely uninhabitable by an European. The excessive heat in which the Canadian lives, within doors, is sufficient to kill anyone, not from his infancy accustomed to that temperature.'[14] The climate in parts of Canada may have been more severe than that in Brittany, but the adaptations and the resources offered a more favourable, not less attractive, environment in which to raise a family. A scholar who undertook a careful study of pioneer society has argued that the Upper Canadian farm family 1860 harvested the requisite supplies of food, possessed sufficient clothing, and lived under adequate shelter. It was likely, he concluded, that 'real poverty' was not widespread.[15] His conclusion mirrors a common refrain from the Canadian prairies, in this instance reported by a child of Lebanese homesteaders in the bush country near Lake Winnipeg during the 1920s: 'Enough to eat, but no money.'[16]

The survival metaphor is misleading. It exaggerates the conflict between pioneer and environment, and it underestimates the achievements, material and aesthetic, of the settlers. They may have produced little or no surplus. They may have saved nothing for their children. But they made ends meet. In fact, they were stronger and healthier than their European contemporaries.

III

The frontier thesis seems similarly threadbare today. It implies that everyone started out on terms of equality, that the fittest survived, and that Canada's institutions of government were made more democratic as a result of the pioneers' protests. Again the easy generalizations about equality disguise the facts of unequal distribution of wealth and unequal access to education and political power that beset these communities.

Elizabeth Goudie's experience contradicts easy generalizations about frontier equality. Her life shows us that some pioneer settlers were very poor and some were not. Goudie wrote of the family's circumstances in the late 1920s: 'As each year went by we were a little more in debt. The Hudson's Bay Company was getting a bit more impatient with us.' After moving back to her father's district, she continued to be concerned about their income. In 1933, she reflected on Horace's fate: 'I looked at my son that night and thought, he will have to go through the same hard life as his father ... Our men and boys had to work in all kinds of weather to keep the ball rolling. When the winters were over we were just about exhausted, but we always managed to pull through some way. We were so glad to see spring again.' Elizabeth con-

fessed that in the leanest year, 1938, the late arrival of a supply ship actually placed the family in jeopardy: 'I was
cooking a pot of seal meat every day to keep Jim and me
alive. The children were so tired of the meat we could hardly
get them to eat any. They would eat fish but not meat ... There
was no one who really starved but some of us were pretty
close to it.' As Elizabeth wrote in the conclusion of her manuscript, 'It was a good life, a very plain life. Just poor people –
most of us were alike, but life didn't seem hard.' She saw
social equality around her, but it was the equality of poverty.

Yet a closer look at the picture that she drew reveals that
Elizabeth's district did indeed contain hints of social inequality. For example, on the other side of Elizabeth's world lived
the mission doctors. They were, as she reports, good people
and good citizens, committed to their charges and in one
memorable case even determined to return to the country for
burial. She wrote of this individual that she 'admired him
very much. He was a good straightforward man and we
could always expect a good handshake from him.' The fur
company officers similarly belonged to a different stratum of
society. They could hope to be promoted to greater responsibilities and rewards, as happened to the most famous of the
Labrador alumni, Donald A. Smith, whose brilliant career
took him to the governorship of the Hudson's Bay Company,
the presidency of the Canadian Pacific Railway, the exploitation of Middle East petroleum, and, by the early twentieth
century, as Lord Strathcona and Mount Royal, the status of
one of the world's richest individuals.

This rough caricature of social inequality on the Labrador
frontier – one that juxtaposes Elizabeth's household with
that of the doctor, or of Donald A. Smith – has been sustained
by recent scholarship. A careful study of two Nova Scotia
districts in the mid-nineteenth century offers a clear illustration of these economic circumstances.[17] It concluded that a
small group of farm families (perhaps one in five) enjoyed
'large, profitable holdings,' a larger middle group (perhaps
two in five) earned 'a modest competence,' and a third

group, (at least one household in three) 'generated less than half their subsistence requirements from their barns and fields.' The last group scrambled for waged labour, hunted and fished endlessly, and relied on informal reciprocal gifts to keep body and soul together. The scholars conclude: 'Life's race was run from a staggered start across far from uniform terrain.'[18]

These northern North American communities had a few prosperous families – merchants and their agents and perhaps a few professionals – and a large number of the poor. Between these two poles was a layer of reasonably secure households. This story repeats itself across the continent. The gap between the very prosperous and the poor was unmistakable; the differing proportions in the middle were noteworthy; but the structure, taken as a whole, offered the prospect of an obvious difference between social groups, not simply a frontier of economic equality.

The myth of the pioneer on the open frontier also carries an implicit assumption about race. W.L. Morton, for example, assumed that the status of progressive entrepreneur in Red River should be bestowed on British newcomers who tried to raise crops and herds. In contrast, a recent social history of the community judges that the truly adventurous spirits of this colony were the buffalo hunters, mainly French Métis, who took greater risks and created a more recognizably 'modern' society based on capitalist principles of markets and trade.[19] These innovative prairie Métis set up family production units for pemmican and buffalo robes between the 1840s and 1870s. All members of the family participated in harvest and production. However, their rapid adjustment to a new economic opportunity – a capitalist enterprise of sorts – came to an abrupt end with the depletion of the natural resource, buffalo.[20] Ironically, early writers within the frontier interpretation never congratulated the Métis adventurer but instead endorsed the culture of the cautious, stolid – and poor – British farmer.

The frontier thesis is also implicitly male-centred and

thereby underestimates the work of women as new settlers. Elizabeth Goudie, eldest child, had to do 'girls' work and boys' work' in her early years. She learned 'men's work' in order to keep the household alive when her husband was out on the trap lines. She recognized differences in role, but she also knew that household needs took precedence over gendered conventions. The French historian of peasant women Martine Segalen has argued that women were so crucial to the household economy as producers of wealth and bearers of children that 'the man–wife relationship [was] based not on the absolute authority of the one over the other but on the complementarity of the two.'[21] None the less, the entire institutional structure of these societies, as well as fundamental household assumptions, favoured the male over the female. St Paul's injunction to women – 'Obey thy husband' – did represent the cultural pattern. As Elizabeth Goudie commented: 'It was the custom for the man to run the home, the women took second place. A woman could have her say around the house but about the main things in life, the man always had his say. His word went for most everything. Women accepted this and thought nothing about it. They were not hard men; they were kind. They were not very hard to please, so a woman could sew and cook and look after her family; that was the most that was expected of the wives. Jim always got me a cup of tea after I had all the children put to bed when he was at home.'[22]

The frontier did not offer 'equal opportunity,' as that term is usually interpreted. Grit and hard work and sheer cussedness were not enough to ensure the accumulation of wealth. Men did enjoy precedence. A significant proportion of first settlers did have it rough. And in the years since, preconceptions about racial character have skewed the story. This is not equality, nor is it the model for democracy.

IV

The staple thesis also distorts the features of early North

America. Despite the careful phrasing of such proponents as Harold Innis, the model that has been extracted from the writings of staple theorists simplifies the past in order to clarify the processes of economic change. As an economic model, it concentrates on the export staple and thus on the moments when human actors participated in the staple's voyage from land to market. The great corporations that organized the harvest or the transportation appeared often in such histories, as did their leading officers, but what of the lowly trapper, fisher, or woodcutter? Given Elizabeth Goudie's story, what of the household, on which the entire enterprise rested? Although Elizabeth and Jim had several good harvests, including a banner season when their furs brought in $1,000, they contributed very little to the export stream in many other years. They worked just as hard in the lean times, but their activities would have figured only sporadically in a history of the fur export business because they often did not produce enough to be included in the Hudson's Bay Company's account book.

There are several aspects to this revision of the staple model. New France, in the accustomed view, existed for the purpose of exporting the staple product, furs. To be sure, the fur trade flourished in the century after 1660, sustained a small bourgeoisie, and employed a small number of casual labourers. Nevertheless, New France was also a farming community. The development of its seigneuries was largely independent of the fur trade and far more successful in the production of goods. The rate of growth of the agricultural economy from 1695 to 1740 was comparable to that in the American colonies.[23]

At one time, historians would have stopped their discussion of the habitant at this point. As R. Cole Harris had then concluded, 'Farm families lived in rough sufficiency, their lives dominated by the seasonal rhythm of the land, not by the more powerful people who lived in other ways ... With the safety valve of such [cheap] land, an egalitarian, family-centred, rural society would be able to reproduce and extend itself generation after generation.'[24] The limits on resources and the changes in international circumstance affected the habitant household, however, as they did the Métis and the Atlantic village. The supply of land in the immediate vicinity was not elastic. Moreover, the seigneur imposed taxes; several scholars estimate that their various levies amounted to 10 to 14 per cent of farm income – a sum nearly equivalent to the total 'net income,' or gain, in a year.[25] By permitting the seigneur to skim off the family's surplus, the land tenure system doomed habitants to labouring on increasingly small, unproductive farmsteads.

A crisis in this once-closed system arrived in the early nineteenth century, again because of limits to the farm's productive capacity and changes in global circumstances. Crop diseases struck, yields dropped, and the levies remained. For some peasant families, wage labour in the fur trade or the lumber camps and timber drives became necessary to pay the seigneur's taxes and to ensure the household's survival. By the 1830s and 1840s, the fur trade, the timber trade, and the wheat trade may all have supplied portions of the household income, along with the garden and wood plot on the seigneurial holdings. Like many modern farmers throughout the world, a stratum that had once enjoyed a reasonable and relatively independent existence became semi-proletarian, resorting to wage labour to reinforce members' single greatest possession – the farm itself. Thus the dominant reality in the economy of nineteenth-century French Canada was not the staple trade alone but rather the *combination* of labour in the staple enterprises and work on the farm.[26]

As the backbone of a growing economy in the Maritime colonies and Upper Canada (later Canada West and eventually Ontario), farm households in those jurisdictions too generated surpluses. Such profits, according to economic theory, eventually could be invested in infrastructure such as railways and manufacturing plants and thereby underwrite the launching of a modern economy. This is a version of Innis's staple thesis.[27] In the first half of the nineteenth century, the proportion of the total population of English-speaking northern North America embraced by the category of farm family was very large, including over half of Ontarians and a substantial number of people in the Atlantic colonies who combined farm, fishery, and work in the woods. They shared with the Métis, the habitants, and fishing villages a family-based strategy of economic production and inter-generational transfer, as well as reliance on the fruits of the land. Yet, if one took the staples model literally, one might expect to find that many farms shipped grain to international markets soon after clearing timber and that farm families subordinated all other activities to that end.

Ironically, the Goudie family would not have made it into discussions of farm staple production, any more than into histories of the fur staple, because it produced no surplus for export. Nor would thousands of other families that combined harvesting from fishery, trapline, forest, and farm to make ends meet. Indeed, one study of Canada West in 1861 concluded that only one farm in six produced significant surpluses of farm goods for local or international markets in that year.[28] These exporters, usually owners of fine estates 'on the front,' near Lakes Ontario and Erie or on the main rivers, were the largest farms, the closest to transport routes, and the longest-settled.[29]

V

Any new interpretation of the past of northern North America must be based on the new 'social history' approaches to the people's economy. Its themes must be barely adequate income, intergenerational stability, and exploitation of common people within a larger mercantile exchange system. It should not reject the idea of a common national or northern North American experience that was such a powerful force in the historical writing of the preceding generation. It should propose that hunter, farmer, habitant, and fisher households had much in common. However, it must suggest that the old economic themes are not as sweeping and conclusive for modern readers as they were for a previous generation. Genealogy, survival, frontier, and staple must make room for family and gender history, on the one hand, and for inequality and exploitation, on the other.

How do today's social historians revise the previous generation's interpretations? In the words of a recent textbook, they say that pioneer communities were dominated by 'inequality, insecurity, and mobility.'[30] They would agree too that one finds the decisive moments of life – a sketch of the lives of everypersons – in Elizabeth Goudie's memoir: these moments include memories of a child's work in the household and of relations with parents; of courtship, marriage, births; and of daily and seasonal routine that became engraved in one's being. I suggest, too, that social historians perceive the settlers' universe as one wherein illness and hunger and insecurity and fear are balanced by satisfaction and comfort and achievement.

Each of the settler households produced a significant proportion of its subsistence needs. Each used the labour of the household. Each relied on family relationships to establish the fundamental lines of labour discipline. For these reasons, each household possessed a degree of independence from the wider world through its relations with the land and its ownership of production equipment.

Yet each household also produced commodities for the market and relied on merchants' links to wider markets for the continuation of the system. Like many commodity producers around the world, these households were exploited by merchant traders and were always reliant on nature's bounty – not merely staple exporters, not reliant only on frontier qualities, not completely constrained by the sheer struggle to survive, but definitely locked into a system that appeared to offer independence, while, beyond their sight and their control, it limited their economic and cultural choices and demanded their obedience in ways that they could not even imagine.

Does this short outline of filiopietist, frontier, staple, and survival histories suggest that such interpretations might speak eloquently to multicultural classes of students in inner-city Toronto or lower mainland British Columbia or Halifax today? Surely not. Instead, the critique that parallels the outline reveals how times and perspectives have changed. However, the new social history conclusions do not distinguish northern North America from the rest of the world with sufficient clarity to identify a distinct Canadian heritage. In this sense, the national critics of the new social history, such as J.L. Granatstein, who have publicly bewailed the failure of this version of the past to come to grips with national questions, do have a point. To respond to this critique, one must turn from economic history and genealogy to culture. One must come to terms with communication and politics.

Elizabeth, like all the others in this society, was a jack-of-all-trades: she carried out all the tasks of a woman and, when necessary, those of a man as well. She wrote: 'Life was pretty rugged for a girl in Labrador. In my day there was not much around except lots of hard work. We didn't get much time for play. If we went out of the house we had to take the baby and look after it when we were playing so we were never free from some kind of work.' Her remarkable memoir serves as a history of a family, but it also opens a window into another dimension of time, space, and political possibility.

4

Family Chains and Thunder Gusts

If the usual interpretations of settler history no longer knit the present to the past, they must be revised. Yet older citizens often trot out yesterday's hard-won truths when they explain to young people why community-wide issues deserve continued attention: your foremothers and forefathers endured hardship so that you might find life easier; your ancestors opposed autocrats in order to build democratic institutions; your predecessors built a transcontinental nation on the transportation systems of the time, proving that adaptations to changes in communication must take the community interest into account. It is easy to see survival, frontier, and staple interpretations in such invocations, but do the stories move today's ungrateful wretches to social and political commitment? If not, does the fault lie with the ingrates or with the stories they are told?

This chapter considers a different argument for continuity between past and present. It proposes that the family chain is a powerful metaphor for a new version of time and space. It suggests that striking changes in life's dimensions were wrought by the slow spread of literacy and print. It also proposes that we should see the politics of the eighteenth and nineteenth centuries as a succession of thunder gusts and that even such seemingly ephemeral outbursts affected the evolution of the community. This was a cultural age different from that of the pre-contact Aboriginal societies, but different too from the insistent rhythms imposed by the machinery of mass production and mass communication after the Industrial Revolution.

Through Elizabeth Goudie's memoir in particular, I look in this chapter at how the specific dimensions of, first, space and, second, time affected the development of northern North America during the age of textual societies. Each local community was built on household economic units. Each relied on literate as much as on oral communication – my third theme here – yet each was also closer to its Aboriginal roots than popular myths have indicated. Fourth, each community adjusted through political activity, understood broadly, to revolutionary changes in communication, although each remained a peripheral economy and each expressed important meanings in religious language.

I

One of the saddest experiences in Elizabeth Goudie's life was the death, perhaps by gastrointestinal disease, of Robert Bruce, the third child whom she had nursed so ably six months earlier in the wake of a cabin fire. She writes precisely of the events that followed his death: of Jim's building a platform to hold the body above the ground for a few days, during which time the men made a casket and laboured to dig a grave in the frozen ground; of her caring for the other children, who had contracted the same disease; and of the funeral service: 'The third day we took the little corpse with us. The only minister around was a long way away. There was an old English gentleman who lived near us who could have buried him for us, but at that time he was in bed ... and could not walk. We took the body into his house and he read a part of the burial ceremony for us. We went on to the grave about three o'clock in the afternoon and Jim and I had to finish burying him ourselves. The two men we had with us could not read and Archie [Jim's brother, who was partly responsible for the disease's reaching their household] said he could not bury him. We had no other choice but to do it ourselves. It was one of the hardest tasks I did in all my married life.'[1] Time passed, the other children recovered, and the tasks of the changing seasons filled their days: 'But the family chain was broken by the death of our little one.'

The family chain is a striking image, but the notion that it had broken forever is probably misleading. Elizabeth had to deal with the tragic death in a struggle that encompassed the rest of her life. By doing so, and by writing so honestly in her memoir, she was able to cast light both on her own sense of life's meaning and on her wishes for her grandchildren. In these messages one discovers what might constitute another founding influence in today's Canada. To grasp Elizabeth's use of the term 'family chain' is to appreciate her vision of a home and its history. She was tied to this earth by her sense of place – that is, the location of specific households in specific spaces – and by her sense of time, in which her family represented a link in a longer chain that was rooted in a Christian, Bible-defined universe.

What can one say about Elizabeth's sense of the place in which she lived? The natural world must have been as real and vital in her daily life as it was for those who grew up in Aboriginal societies. When she was twenty, a young mother, she and Jim and their baby travelled south by canoe to the settlement of Hopedale. The second day (of three) on the Atlantic was nerve-wracking: they crossed a wide and potentially dangerous bay in their fourteen-foot Peterborough canoe, Elizabeth nursing the baby and then paddling while the baby slept. Her relief on land that evening was almost palpable when, after a good meal and a quiet evening, baby and husband fell asleep in the tent. Although she wrote her report forty years later, it still carried emotion: 'I could not get to sleep. I lay awhile in the tent and then I went outside. It was just beginning to get light. It was about one o'clock now. I went for a long walk. The sky was so beautiful and clear and the water was so calm that I was just lost in the beauty of everything around me. The land was rugged and bare, but it was beautiful. The little birds were waking up for the morning. The squirrels and rabbits were pouncing about the side of the hill. I looked up and the sun was rising. I thought I had better get back to the camp. When I had got back it was two o'clock. I was away an hour but it had only seemed about ten minutes. Jim and the baby were still asleep. I lay down and fell asleep myself.'

Six years later, after the birth of Robert Bruce, during late autumn and early winter, when Jim was away on the trap line, Elizabeth again was struck by the natural world around her. In her scribbler, she recalled: 'By November it started to get pretty cold. We had a couple of snowfalls and the ice gathered on the shore. I used to look out late at night and there would be a breeze blowing over the salt water. It showed the blue and green shades in the rippling sea tops. I would go outside after supper when I was alone and take a long look at the beautiful northern lights dancing across the sky in their beautiful shades of light yellow, purple and orange. This was all I had for entertainment but I loved it. I often looked at the northern lights for an hour when my children were asleep. Sometimes, during the day I could look out and see a seal or a duck swimming by.' Hers was a lonely existence, as Elizabeth makes plain in such passages, but nature was partner as well as opponent. Like the Aboriginal people, Elizabeth Goudie belonged to the land more than the land ever did to her.

Elizabeth learned about her world from her parents and from her husband, who was an able hunter. After the dangerous day during their canoe trip along the Atlantic coast, when she had been so worried about herself and her infant son, 'Jim asked me if I was afraid ... I said I was very much afraid. He told me he knew by the sunrise that morning it was going to be a calm day. He went on and explained some things about his travels to me which I didn't understand. He told me that the trappers learned to travel by the signs in the sky. They knew weather was going to blow or be calm by the colour of the sunrise and clouds in the sky ... From the experience that I went through that day I saw that there was a lot to learn.' Later, Elizabeth compared her father's woodlore to that of her husband. She recognized that Jim 'had a better knowledge of how to build a fire and set up a camp for the night. He was such a lover of nature that he knew just how to handle it. After I watched him a few times and saw how he managed things so well, I never felt afraid with him in a canoe or boat or on dog team.'

The dimensions of space and time that shaped Elizabeth's perceptions of life were closer to those of her Aboriginal neighbours than to those of Canadian citydwellers today. Therein lies one of the values of her narrative and an illustration of why we can consider her as both a settler and the last of the pioneers. The rhythms of her activity were similar to those of pre-contact Aboriginal people, and her reactions to daily circumstance were probably very similar to those of her Aboriginal neighbours. Her stories echo the seasonal round that figured so prominently in the story told by grandmother Andre.[2]

In the conclusion to her narrative, Elizabeth linked her life with the space in which she had passed it:

> I am very proud of my country, Labrador. That name goes very deep within my being, the beauty of its rivers and lakes and the beautiful green forests ... I look out over the miles and miles of hillside untouched by man and I wonder how much longer are we going to be able to keep its beauty. The name Labrador holds something hard to explain but I would like to explain it in my own way, and that is peace – a deep peace within that helps to make all its hard work easier to take ... Now the days of the trappers are behind us but I am proud I had a chance to live such a life. The wife of a trapper played a great part in it because she had to live as a man five months of the year. She had to use her gun and she had to know how to set nets to catch fish and also how to judge the right distance from the shore to cut her fish holes because If you cut them in

too shoal water you would have to move out into deeper water and then you would have to cut another hole in three-and-a-half feet of ice. I learned all of this the hard way ...

Labrador ... Its people were a *people* in the past. The trappers made their own laws, they respected each other's laws and they carried them out to the best of their ability. I will never change deep within my heart and I hope I can be a friend to people. We should all strive to live in peace with one another. That's the only way to live right.

The sequence of her thoughts requires particular attention: it traces a line from Labrador as place of physical attractions to one of personal peace. She then outlines her work and its special qualities that demanded a certain kind of learning and that provided a sense of achievement – achievement that she earned 'the hard way.' And *then* she turns to her sense of Labrador as community – a distinct community – with its own laws. Finally, she identifies the ideal life as one in which friendship and peace prevail. In brief, she sees place, work, a sense of belonging to a community, and learning to be at peace with oneself as a single chain.

It is the rootedness of the passage that strikes home. Elizabeth belongs in a particular place, in a land that she knows with every fibre of her being. The land offers her a perspective on all of life and all of the universe.

II

Gerald Pocius, a student of Newfoundland folklore, concluded that space was the 'central organizing feature of daily life' in the Newfoundland village where he had lived and worked. I wonder if instead perceptions of space and time do not always operate in tandem in a culture. Elizabeth's sense of family, like her sense of place, gave her a location in history because it placed her as a link in a chain that extended over two hundred years in this part of the world and on a much longer chain that was dimly perceived but none the less real.

In the opening sequence of her memoir, Elizabeth set down the details of her family genealogy. She noted after several statements, 'according to my parents.' The one anecdote that she tells in this section, concerning her grandfather's death, ends with the attribution: 'This little note was handed down from our forefathers.' Her perception of time in this chapter is noteworthy because it is so clearly rooted in a multi-generational vision. Pocius had observed about Calvert, the village he studied: 'No narratives dealing with the homeland survive.' Thus, he concluded, there were only three parts to the historic past: the events of one's own lifetime; the moment of 'firsts' (first settler, first ancestor in this district), as recorded and handed down by one's family; and the period stretching between these two poles, which is a story partly of 'the family chain' and partly of the larger community.[3]

This is not the case for Elizabeth. The wider world, as we can discern it in the parts of her memoir dealing with the years before 1940, provided several narratives linked to a much longer chain. In particular, Elizabeth viewed the world in terms of stories drawn from biblical history and British civilization. Her English ancestor in Labrador, she reported, taught the Inuit orphan to say the Lord's Prayer before he married the young woman. Her parents read the Bible aloud daily ('the last thing that was done at night'), observed the Sabbath, and followed the commandments of Moses ('Thou shalt not - so to speak,' as Elizabeth observed). Her husband, Jim, wanted to name their third child Robert Bruce: 'He had read a book about Robert the Bruce of Scotland. He had been a great man. Jim had Scottish blood in him. He said that they were good, straightforward people. That was how he felt about them. I joshed with him and said, "You and your Scotchmen."'

The depth of the Christian assumptions in Elizabeth's life becomes apparent in brief and admittedly rare expressions in the memoir. Of the month after Bruce was so severely burned, when she and Jim worked feverishly and tirelessly to save him and to keep the family going, she wrote: 'Many, many times after, we both wondered how we had done it. But with God's help, we fought for his life together.' Several years later, after Jim's trap line accident, when he had recovered sufficiently to speak, he said to Elizabeth, 'I knew God would bring me home to you and the children.' She added: 'It made me wonder if it had been his faith that kept him going.' In the many days alone in the cabin, when the children called for their absent father, she sang 'this little song that I made up: "Hush my dear, lie still and slumber, Holy angel guard thy bed; Heavenly blessings without number, gently falling on thy head."' Reflecting on her husband's life, she wrote: 'He never worked on Sunday unless he had to. He kept that day for rest.'

As the comment about Sundays suggests, Elizabeth's beliefs were linked to a pattern of British customs associated with the Christian calendar. Of a rare Christmas when all the family was home, she wrote that they spent a very quiet day:

> The children hung up their stockings in those days. There were not many gifts. The boys had a rubber ball each and our little girl had a little doll and a few candies that had been saved in the fall for Christmas. There was an old man and his wife living near us and we took the children over to see them in the afternoon. We sang Christmas carols and read the Christmas story. We went home and had a quiet evening by ourselves. This was the way things went at Christmas in Labrador in the bays on the coast. We always visited with one or two families. The Christmas feast was made up of fresh baked partridges or baked rabbits or caribou steaks. The pastries were partridge-berry pies and the cake was made of molasses, raisins, currants, spices and baked in the oven like you bake a plain white bread.

Thus Elizabeth's sense of the wider world – its dimensions in space and time – relied on a biblical vision, a smattering of Old Country literature, and a Christian festival calendar.

These few hints imply that there were at least four kinds of time in Elizabeth's daily life. She recognized a version of historic time that differed from the pre-contact Aboriginal world-view; this dimension included biblical history and integrated stories from the Old Country, including a mention of Ambrose Fisher's ancestors and of Robert Bruce. There was also ecological time – the recurring cycle of seasons and resource harvests. There was as well structural time, marked by the stages of a life and the milestones such as births and marriages and deaths. And then she had a sense of the family chain and of the extraordinary departures such as the transformation of Labrador after 1939 that she witnessed, both of which belonged to the same historic time as the Bible but differed from it because they were located in the only place of which she had direct personal experience – *this place.*

The world of the settlers differed from that of pre-contact Aboriginal people in several aspects of space and time. The dimensions of the Aboriginal world found expression in the language or forms of nature itself – for example, the distance of voyages estimated in days, the passage of historic time recorded in the annual images of the winter count, a person's entrance into the world of frogs or foxes by means of contemplation and dreaming. The dimensions of Elizabeth's land were measured by miles and years and hours on the clock, as well as by days of travel or seasons of experience. Moreover, her calendar belonged to everyone in the community, literate and illiterate, because written communication had transformed such perceptions in a uniform way that transcended local places and spaces.

After the arrival of literacy and print, the settlers and their Aboriginal neighbours alike were absorbed into a textual society in which 'alphabetic-letterpress-print' abstractions communicated messages in uniform format.[4] Thus even descriptions of very different terrain and experience – how far, what duration, how much wealth – could be conveyed by a common, print-encased language of measurement and list and definition. Certainly Elizabeth's grasp of world history differed from traditional Aboriginal perceptions. Thus she referred to stories in the Bible, to Ambrose Fisher's arrival in Labrador, and to her own peripheral relation to two world wars in the twentieth century.

Whereas Aboriginal cultures perceived a seamless sphere encompassing the natural and human universe, Elizabeth Goudie saw the relations between humankind and nature in a biblical as well as a nature-shaped perspective: nature and humanity designated separable, though connected, spheres. Elizabeth acknowledged that nature could be beautiful and terrible; but she also believed that human life evolved according to the rhythms and meanings set out in Christian teaching.

One occasion when her conviction slipped was when a lynx attacked her and she faced death. In that moment, as the big cat raced across the frozen bay towards her, she knew utter terror. Six months later, she gave birth to a daughter, another crisis in the life of a household:

> I was afraid when my baby was born that it would be deformed in some way or other because I had such a fright with the lynx ... When the midwife brought her to me I looked at her and she looked more like a lynx than a baby. But it was only her head, her body was all right. Her face was not quite right and her hair was sticking up all over her head. I was so worried about it. She reminded me of the lynx because when I had looked at it coming for me out on the ice, its hair was sticking up all over its head, with two big teeth in front of its wide-open mouth. The next morning when the midwife bathed my baby, I saw there were two teeth sticking out of her gums so I did not know what to expect of her. What if I would have to look at her head with the hair sticking up all her life and perhaps also a mouthful of big teeth?

The association of animal, human, and magic worlds – as opposed to Christian faith – seems very close in this episode, but Elizabeth's sense of humour smooths it away as she retells what is no doubt one of the family's best-loved funny stories.

The dimensions of space and time in Elizabeth's memoir reflect both a traditional and a textual approach to such matters. This was the world-view of immigrant settlers in northern North America. She, and they, expressed space in terms of particular known places and measured it by the natural activities of walking and canoeing, as well as by such textual abstractions as miles and minutes. She, and they, measured time by links in the family chain as well as by the Bible.

III

The world of the settlers was by definition a textual society, one in which print established the communication context, although many of the settlers would have had only a passing acquaintance with the daily business of literacy-driven institutions and perhaps with the formal skills of reading and writing. It was also a world in which commodities themselves – staples such as fish, fur, timber, and wheat – acted as vehicles of international communication. Like so many others who produced staples for export, the Goudies discovered that their income was often very small in relation to the labour expended and that they had no control over the price of the goods that they sold. As in Hugh Brody's story of decision making among Aboriginal people, so with settlers, the *system* of trade and the nature of communication shaped the ordinary person's reactions. Their responses, most of them oral rather than written, constitute the community choices that we label 'politics.'

It would not be accurate to distinguish too sharply between the two kinds of social communication – the oral and the literate – as if they were mutually exclusive. Rather, within the settler communities of northern North America, the two existed together, creating a textual universe in which everyone dwelt, including those who neither read nor wrote. Writing made it possible for the elites, as a student of ancient

Greece suggested, 'to facilitate communication over long distances and therefore control over larger areas, to make commands more authoritative and unchangeable, and keep records of large numbers of people and taxes. One could add a further function, the use of writing for display or propaganda, which deserves distinguishing from the use of writing for administration.' This does not mean that writing eliminated 'oral modes of proof, memorial and communication.'[5] Instead it permitted the simultaneous employment of the two modes, although which was authoritative in a given moment and community depended on the activities involved.

To understand the relations between the two – the oral and the literate – and to see the intersection between mode of communication and political sphere, consider again the experience of the Goudie family.

Elizabeth 'had the privilege of going to school two winters ... I spent three years in school in Mud Lake but really I had about four years in school if it were all put together. I got as far as the fourth grade in the old English readers and the Kirkland Scott Arithmetic Book.' Her husband, Jim, 'never went to school in his life but he learned to read and he could look after his own affairs. He loved to read.' In the year before the arrival of her first child, Elizabeth 'had lots of time on my hands. I would go hunting and fishing and in the long evenings I would read a book if I had one.' In their family years, however, they never had time for such activities. Her two eldest children had little more exposure to books: 'Horace and Marie both went to school for a while. Horace was only eleven years old when he started going out into the country on his trapping lines. He only had about three years of school. Marie was kept in school longer ... Horace wanted to go with his father so much that I let him go ... He would go to school in the fall and go into the country in the spring with his father ... Jim was beginning to find the work hard and Horace could help him with camping in the evening. He was good company for him so I never tried to keep him home after that. I saw what he meant to his father, as Horace grew older. When Horace could stand the hard work, he went with him all year round.' These are statements about the formal arts of literacy and the inescapable necessities of life; they demonstrate that the Goudie family understood the value of books and print but, in the years before 1939, rarely had the opportunity to work within the world created by print except through the Bible, their domestic accounts, and a few randomly selected volumes of literature.

Communication is much more than simply a matter of print and oral messages. It also comprises contact between individuals, the expression of understanding or sympathy, and even the assurance that the support of others will be forthcoming in time of need. Elizabeth noted that one of the great drawbacks to their northerly location in her early adult years was the communication system: 'but what bothered me most was that I could only get letters about twice a winter from my family back home.' She added that, with the passage of several more years, she felt the isolation and the poverty more keenly: 'I missed my parents as I had been away from them now for four years. We only got letters from them about twice a winter and about three times in the summer. We were not making money to spare and there was no way of getting home for a trip to see them. Mail was delivered in winter by dog team and in summer by a steamer from St. John's. As each year went by we were a little more in debt. The Hudson's Bay Company was getting a bit more impatient with us. In 1928, when we went to get our food again for the winter, the Hudson's Bay manager told us that if we could not pay off our bills that year, he would have to cut off our credit.'

In this passage, Elizabeth linked communication, financial solvency, and family relations. The reason for the association among these three spheres was the decision that followed: she and Jim concluded that they could not make a go of things in the district where they had been settled for six years. They would instead return to her parents' territory and assess their chances of settling in that part of Labrador. Loneliness and sadness are part of the equation (Bruce had died just a few months earlier), but so too are the related needs for everyday economic survival, a reliable source of emergency supplies for the household, and communication itself.

When they concluded in 1929 that they could not make a living in the northern districts, she and Jim returned to Hamilton Bay: 'We hadn't told father how poor we were, but after a couple of weeks, Jim said we would have to tell him. Mother and Father were alone one evening, so we told them. Father felt awfully bad. He did not say very much at the time. The next day he talked it over with my brothers and agreed to take us in for the winter. We were so thankful for that.' Again, the family as production and consumption unit is central to the story. In times of dire need, a family moral economy could swing into action.

Several years later, in 1935–6, Jim sustained the axe wound that nearly took his life and left him in the hospital for many weeks. Horace, then fourteen, could not keep the family going, as Elizabeth noted in an understated passage: 'I still had to go to the Government. They gave only six cents a day per person. We got flour, tea, molasses, a few pounds of butter and a few pounds of dried beans and peas. The children didn't have much to eat that spring. When Horace sold the few furs he had, I bought a few tins of milk and some rolled oats for their breakfast and managed to save five dollars for the midwife when she delivered my baby.'[6] This paragraph, with its reference to 'the Government,' is the closest Elizabeth's memoir ever comes to a mention of formal political affairs.[7] What is one to make of such social welfare in the history of the Goudie family of Labrador? Is this an episode in the long narrative of mere survival? An expression of frontier democracy? An exception that does not fit the model of staple economics? Or something more?

IV

Given the Goudies' remoteness from other families, political activity – parties, platforms, campaigns for reform – was probably not a part of their life before 1940. Yet the issues associated with their survival might seem 'political,' in its broader sense. Elizabeth and Jim had undertaken in 1922 to build their married life in a new fur trade district, and, although she presented no reason for this strategic decision in her memoir, it is possible that the fur company had encouraged what was in effect an experiment. The family had laboured long and hard to make ends meet in the new territory. After seven years and considerable investments in housing and furniture, as well as careful study of the land and its resources, Elizabeth and Jim had abandoned their accumulated 'capital' because the impatient company and its manager believed that their family enterprise – their experiment – was not profitable.

Should the rule of profit and loss, as defined by the company, prevail without appeal? Did the Goudies deserve no compensation for the investments they had made? What of the prices the company was offering for their furs and what of the company's profit levels? Should there have been a reassessment of the relative revenue shares allocated to investors, traders, and trappers?

Jim and Elizabeth would probably not have dreamed of such defences of their effort and of the possibility that a larger economic and political system might have borne some of the responsibility for their circumstances. Their acquiescence seems to have been the natural response of many settlers in all the societies of northern North America. It seems entirely reasonable to us today too, because the popularity of survival, frontier, and staple interpretations channels and limits our thinking. But the nature of the political response of the common people has more to do with the communication context, less with these time-honoured frameworks. The Goudies, like so many settlers, lived in relative isolation. Because of the communication system within which they operated, they would have had little opportunity to meet with their compatriots and to 'compare notes.'

When the circumstances did permit greater community cohesion, the most likely political responses among settlers can be itemized as petition, sabotage, and popular revolt. No long-term, organized, cooperative attempts to restructure the economy or to redistribute public power in the interests of 'the people' appear in the political history of the seventeenth, eighteenth, or early nineteenth century. Such movements and parties awaited broader democratic change. As Goody and Watt have suggested, 'Democracy as we know it is from the beginning associated with widespread literacy.'[8] Thus improved popular communication networks and sustained democratic political protest movements went hand in hand.[9] The prairie Métis, the Newfoundland fishing families, the habitants of Quebec, and the pioneer farmers in other parts of northern North America were limited in their power to effect change both by the rules laid down by the established order and by their inability to contact, debate with, and mobilize their brothers and sisters.

The political world of the habitant household, for example, was quite confined. Yet, unlike the Goudie family, habitants did participate in moments of popular outburst. In seventeenth- and eighteenth-century New France, 'demonstrations were the only collective means by which the habitants and lower classes could influence those in authority.' Marches and demonstrations occurred when the price of bread or salt, the imposition of a labour duty, or the reorganization of a church parish aroused the people. These 'Thunder Gusts' did 'more Good than Harm' because they offered a means of collective expression and warned officials that a grievance existed. They were spontaneous, had limited targets, were generally not violent, and, rather than seeking new powers or drastic change, were defensive in the cause that they espoused.[10]

The Métis first organized as a collective force to resist the new Red River Settlement in 1815–16. They rode through the crops of the recently arrived Scots, threatened violence, and eventually fought a pitched battle to assert their claims. A generation later, a Métis mob's threat of force at the 1849 trial of one of its number, Guillaume Sayer, won official recognition of the Métis right to pursue the fur trade in the face of an official monopoly held by the chartered trading enterprise, the Hudson's Bay Company. The failure of petitions to bring about reforms in 1869 and 1884 precipitated violent uprisings that the Métis insist were movements of resistance rather than 'rebellion.' In every case, public action was organized by word of mouth and founded on community consensus. The role of Louis Riel in the events of 1869–70 and 1884–5 is instructive. He was both the orator and the negotiator. He won support by speaking but also by writing. He was relied on partly because he could deal with the new literate world on its own terms as well as with the oral world of the Métis family. He knew both the law of constitutions and the magical foresight of the hunter. He was a prophet as well as a politician.[11]

Newfoundland villages might seem more enclosed, more garrison-like, than the Métis settlements. They were none the less based on extended families, like their prairie counterparts, and they were similarly subject to the power of a merchant trader. Villages were sometimes rife with tensions. Quarrels over the distribution of resources and the burdens of work could become bitterly divisive. Distinctive local customs helped to resolve such problems. Christmas 'mummering,' when people dressed up in disguise to visit and to party, aided in the reconstruction of personal alliances; the act of stealing foodstuffs from one household and presenting the stolen meal to an extended village group – 'scoffing' – expressed intimate antagonism while reaffirming the villagers' community membership. Gerald Sider, an anthropologist, has argued that such events expressed the system's contradictions, facilitated its continuation or 'reproduction,' and marked the village off 'as a special, and in a collective sense private, place.'[12] In Sider's interpretation, the distinctive customs enabled the people to acquiesce in – to justify – their being exploited while expressing anger at their powerlessness.[13] Other scholars have challenged this picture of collective weakness. They have pointed to popular violence, concerted protests, and strikes as evidence that the fishing families did fight against the system that exploited them.[14]

Religious expression also played a part in the settlers' political history. The camp meeting of the late eighteenth and early nineteenth centuries – a form of religious revival – represented a dramatic illustration of *political* experience. The character of such meetings has been described as a 'theology of feelings,' and a 'religion of experience'; both phrases emphasize that such religious events were guided by emotion as much as by reason and were affected by a battle within the individual between sin and salvation.[15] The camp meeting was a cultural form accessible to the literate and the unlettered alike. The rough schoolhouse, the community hall, or the clearing in the bush would resound with shouts and cries, at times in unknown languages; neighbours and children and husbands and wives could become heated, ecstatic, frenzied, sometimes uncontrollable, driven by emotion and tears and agony rooted in some unknown force. These camp meetings, which were mass events involving a substantial portion of the community, lasted for several days and involved continuous preaching, singing and exhortations read from the Bible. Thugs and other rough elements would be excluded to ensure that worship was not disrupted. Words burned like 'fire,' God's intervention brought 'showers of grace,' and eventually a sinner fell 'like a bullock at the slaughter.' William Westfall has summarized the meeting's impact: 'The preaching was intense and emotional; the texts were drawn from some of the most highly charged passages of scripture. In the language of revivals, the preachers had to "preach Christ crucified"; they had to "preach for a verdict," not to inform and instruct but to bring about an immediate conversion to Christ ... As preacher followed preacher, as exhorters moved through the crowds, as the converted turned upon the unrepentant but wavering sinner, individuals would finally break under the weight of the revival, acknowledge their sinfulness, and accept God's saving grace.'[16]

This cultural form combined written truths and voiced emotions. Elizabeth Goudie was probably accustomed to Moravian worship in Labrador, which was more austere, but many settlers in earlier times would have experienced the force of Methodist and other similar religions of passion. At their heart was the power not just of the text but of the *spoken* word, of preaching: loud and soft, angry and gentle, but always, when done well, intoxicating. This was theatre, a powerful and accessible art in pioneer farm society. Words tumbled from the great orators, as when Freeborn Garrettson drove away opponents from an evangelical meeting in Shelburne, Nova Scotia, in 1785: 'Whilst I was preaching to near four hundred people by candlelight, they ["the devil and his children"] were beating underneath [the meeting house] ... In the midst of my preaching I cried out, *Without are dogs, sorcerers, whore mongers, idolaters, and whoever loveth and maketh a lie.* The company ran off with a hideous yelling, and we were left to worship God peaceably.'[17]

A religion built on individual conversion, despite appearances, need not be private and isolating. Behind the emphasis on the individual lay an equally vital concern for the community, especially for its transformation into a more Christly place.[18] The camp meeting provided the occasion in which community, especially family tensions, could be worked out.[19]

The settlers eventually confronted a larger world, which required further adaptations. The Industrial Revolution and capitalist economic assumptions reached into every corner of northern North America during the nineteenth century. One of the most difficult aspects of these changing times concerned the passing of property and power from one generation of settlers to the next. If the farm, the centre of the family economy, was no longer sufficient to sustain all the children, how should the parents prepare their offspring for the changing world? Elizabeth Goudie, writing in the early 1960s, clearly recognized that some of her children lacked one kind of experience that they would need in the new age

– schooling. Yet she had not been so concerned about the briefness of her own period in school in the opening decades of this century. Why the difference?

The story of Ontario schools offers an illustration. In the mid-nineteenth century, Ontario (then Canada West) built an extensive system of public schools. It was a gigantic undertaking. A new department replaced the haphazard patchwork of local and religious schools and the uncertain arrangements for children's education provided by parents. It required establishment of a new profession (teaching), creation of an administrative network, elaboration of a new level of political institutions, selection of instructional materials, and physical construction of a vast network of classrooms. At the heart of the new system was a determination that the next generation be literate.

Scholars debate the reason for the change. Did it originate with business leaders who sought obedient workers and well-informed consumers? Was it driven by church and social reformers who wanted to build a more orderly world wherein missionary evangelism might prevail? Was it led by the state, which was desirous of central control over governable citizens in order to ensure political stability? Or was the schooling movement pushed by parents who perceived that the economic context of pioneer agriculture was changing and that the new age would require different skills? Although each of these interpretations may have relevance, the last is especially useful. Many poor farm-based parents could not offer land and equipment to all their offspring as the latter reached the age of majority. However, they could ensure that all their children possessed the appropriate intellectual tools. A formal education was the one certain gift that parents could arrange during the adjustment to changing times.[20]

The story of politics in a settler society is the story of movements and outbursts, not of formal political parties. It is also a tale of slow, deliberate adaptation to circumstances that were less than predictable. It encompasses welfare payments, popular uprisings, religious revivals, and the building of schools. And, of course the story is shaped by the means of communication used by community members.

V

As this chapter has shown, Elizabeth Goudie was a link in a family chain. She adapted to drastic change, made ends meet in difficult circumstances, and passed on a proud legacy to her children. For today's Canadians, one of her greatest legacies is her memoir, the result of sustained effort late in life. It provides both illustrations of the lessons taught by the new social history and refutation of the charge that the multiplication of historical subjects divides Canadians from each other. Elizabeth's narrative offers insights into a long epoch in northern North America.

Elizabeth and Jim Goudie, like grandmother Andre, belonged to a founding generation of Canadians. In Harold Innis's terms, the Goudies too were 'fundamental to the growth of Canadian institutions' because they represented a second version of a 'strategy of culture.' These people studied the land's resources, as the Aboriginal people had done, established new family chains on the basis of the land's natural wealth, and intermarried with the first inhabitants. Though they followed Aboriginal example and advice, the Goudies travelled Labrador's rivers and trails with different visions in mind. They represented in their own family story the fact that the two cultural dimensions overlapped in northern North America. Therefore the concept of a country with two histories for two peoples (Aboriginal and non-Aboriginal), as in Australia, is less appropriate for Canada. Canada's Aboriginal and first settler communities folded into each other to some degree (the precise degree is a question worthy of much greater discussion) and lived a common history.

Generalizations about this experience, as expressed by the generation of historians previous to us – Innis and Creighton and Groulx and the others – are powerful tools for helping us understand Canada's past. Settlers anchored family chains, survived terrors and natural challenges, shared in a kind of frontier freedom, exploited potential staple crops, and constructed regional populations that differed in character one from the other. However, these positive statements offer only part of the story as we should reconstruct it today. It is true that these settlers faced periodic crises and constant labour; it is not true that they were in perpetual danger or forever on the brink of death. It is true that some grew prosperous, but many more possessed little property and even less power. It is true that northern North America's regional communities were connected to the expanding capitalist economic system, the 'world-system,' as it has been called, but they existed on the periphery, not at the centre, and this location made all the difference. Noting the distinction between centre and periphery is necessary because the leaders of Europe's empires laid down the rules of the world-system in the great central governments and markets. The people on the periphery of the trading world, like the Goudie family of Labrador, accepted the prices set in Europe as well as most of the other rules that were worked out by such civilizations.

The differences in the characters of the various staples, and the differences between the imperial languages, did not divide irrevocably the various regions and frontiers of northern North America one from another. Common people such as Elizabeth participated in a phase of world history that came to every civilization as it adapted to literacy and print. The natural world shaped ordinary people's responses, acting as a restraint on and an alternative to adoption of clock and calendar, of property law and land sale contract. Thus the 'plain' people of northern North America belonged simultaneously to a literate culture and to a natural environment. They spoke European languages, but their understanding of these new places also was based on Aboriginal teaching.

The circumstances in which they encountered communication technology shaped the ordinary citizen's political experience. Petition, sabotage, and revolt were typical forms of community participation. Christian institutions offered a channel for community experience and often articulated responses to daily circumstance. Only very slowly did literacy come to shape most aspects of their daily expression. Parents still passed on their wealth – so much of it contained in experience rather than in material possessions or written texts – by spoken word and silent deed to the next generation.

Elizabeth Goudie is the representative of this textual-settler society. She created, with her pencils and school scribblers, a work of literature. Her memoir conveys the unity that she perceived between her life and the place in which she lived it. The book illustrates her particular perception of the dimensions of time and space. She did not control the leading media of communication in her epoch, but her steely determination and impressive stamina enabled her to complete a genuine, emotionally precise, written statement of her reactions to the world. Elizabeth's epitaph for her husband establishes this link between place and being:

He was a trapper all his life until a few years before he died when he worked with the Americans as a packer and crater in No. 2 Warehouse at Goose Bay. He trapped a lot of land in this part of Labrador. He trapped on the Hamilton River and to the height of land on Sandgirt Lake ... He did a lot of traveling in his day ... He walked 300 miles when he trapped the height of land near the Quebec border. He travelled miles inside the Churchill Falls when we lived in Mud Lake. When I think of all the hardships he went through for us it brings tears to my eyes so I keep my house open for his children to come and go as they like. I know he would want that so it is going to be that way as long as I can keep it open.

We worked side by side those forty-two years together, and it was pretty rough sometimes. We respected each other and when he was taken from me I didn't feel too bad. Life is meant to be that way. I think a person has nothing to regret when they are happy and we were very happy, so I am quite content

now. There is always something to do and always something
to think about.

The sequence of Elizabeth's thoughts offers valuable clues
not just to her relations with Jim but also to her sense of the
world. She describes Jim's occupation first. Then she turns to
the places to which he belonged by virtue of having travelled
them on foot and having lived on their resources. Next, she
mentions her sense of shared hardships. Finally, she says that
she can be content with her efforts and mentions her children. Note the items on her list: work, character, place, and
family fit side by side in her reflection.

This chapter has presented a multicultural interpretation of
Canadian history because it focuses on an experience – a fundamental transition in perceptions of time and space – that
ordinary people encountered in every civilization. In this
sense, Elizabeth and Jim Goudie offer an illustration of a
powerful and inclusive story. They represent millions of people who experienced this textual-settler version of communication and culture in various corners of the earth between
the twelfth and twentieth centuries. What is significant about
the Goudie family's experience for later generations of Canadians is that Elizabeth and Jim recapitulated the adaptation
in this place and in relation to the traditional societies – represented here by grandmother Andre – that surrounded
them. All Canadians, newly arrived or long-settled, can find
in the Goudies a link between their personal and family
experience, the Canadian community, and world history.

PART THREE

Print-Capitalist National Societies

5

Phyllis Knight and Canada's First Century

The ordinary citizens of northern North America – memoirist Phyllis Knight among them – had to respond to a third dominant construction of time and space in the last half of the nineteenth and the first half of the twentieth centuries. The dimensions of their daily lives were altered by the emergence of waged work in a capitalist society and by an extraordinary succession of technological innovations – the telegraph, the railway, the daily newspaper, the telephone, the photograph, sound recordings, radio, and film, which first appeared and became vehicles of communication between the 1840s and the 1930s. All of a sudden, the message could travel faster than the messenger, as Marshall McLuhan put it. Such an unprecedented combination of changes in work and breakthroughs in communication imposed great strains upon every society, not least on common people. And yet what is remarkable in retrospect is the success of ordinary citizens in bending the economic system to take their concerns into account.

This chapter discusses two major types of historical interpretation – progressive social and conservative political – that have distinguished this new print-capitalist society from its oral-traditional and settler-textual predecessors. It suggests that their often-conflicting historical perspectives on such matters as agricultural life, class, and gender possess less authority in today's context than they once did.

Phyllis and family dog above Second Narrows bridge, Burnaby, c. 1935.

Chapter 6 turns to cultural history to propose a different approach. It concludes that in this age of nations and print-capitalism the common people have also been, to extend Harold Innis's thought, 'fundamental to the growth of Canadian institutions.'

I

Phyllis and Ali Knight can well serve as examples of how Canada's plain people responded to this third construction of time and space.[1] Born Felicitas Golm in Germany in 1901, about the same time as Elizabeth Goudie, Phyllis differed in many respects from her Labrador contemporary. By her own estimation, Phyllis knew more disappointments than triumphs in her life. She never lived without money worries, but she found many moments of happiness in the hard times, according to the memoir that she prepared with her son in 1972–3. She grew up in a poor working-class home in Berlin before and during the First World War, spent many hours at household chores, accepted the first of a long line of wage-labour tasks at fourteen, and contributed much of her salary to her family until she married Ali Krommknecht at twenty-five. She followed her husband to Canada in 1929 (he had gone ahead the previous year), and they worked their way west as the Depression tightened its grip on the country.

Phyllis arrived in Vancouver from Calgary 'with twenty-five cents in my pocket and the soles coming off my shoes.' She pawned their wedding rings – hers 'was a beautiful ring, gold with a little ruby in it; I just loved it. Those pawnbrokers knew how to press people out ... They gave me seven or eight dollars for the rings and with that money I bought myself a pair of sturdy shoes, had a big meal for a quarter, and got myself a room at the YWCA for fifty cents a night.' She then waited on a bench outside the Canadian Pacific Railway (CPR) station – for three days – for her husband. Not having enough money for two tickets, he had joined other men 'riding the rods' from Calgary to Vancouver.

Phyllis and Ali worked on and off for the next decade, sometimes living in jolly settings in tenements shared with dozens of other transient workers, often in poverty, moving constantly from job to job, and sharing what they had with those who needed help. After several miscarriages, they had a son, Rolf, born in 1936 (Ali had changed their name to Knight the year before), and he travelled with them to the timber and mining camps where they found work as cooks. When Rolf was six, and ready to start school, they decided to establish a permanent base. They settled near the water in Vancouver – 'just a shack, a very run down shack, worse than many of those squatters' houses. There were two and a half tiny rooms with no toilet and no basement. It was made from waste lumber and was leaning a bit to one side. The full price was seven hundred and fifty dollars – lot, house and all.' But it had a wonderful view of the harbour and plenty of grounds for a garden. Some years later, with the shack 'literally falling down,' they began to rebuild, section by section 'as we had the money. We tried to do it all with ready cash

Ali, riding the freights, somewhere in the British Columbia
interior, c. 1932.

because we didn't want any big debts over our heads. Only once did we borrow three thousand dollars for the space of a year and a half. I hated to pay the interest.'

Phyllis described the years from 1950 to 1964 as the period when she was a 'grass widow.' Ali worked as a cook either on ships or in camps while she budgeted as carefully as she could in Vancouver: 'We always had to have some savings for when Ali wasn't working. That saving became so engrained in me that I did it even when I shouldn't have. That's why we were never able to get away on a trip or do anything ... You can always find reasons, and they were real enough. We had to pay off the house, or a car, or Ali couldn't go because another job came up. There was just no chance for any family life. When Ali was back in Vancouver, instead of enjoying himself and resting and going out with his family, he was constantly running around for the next job.'

Ali died in 1965 as a consequence of a brain tumour that had been misdiagnosed as a stroke, a medical error that troubled Phyllis deeply. She soon moved out of the house, living first in a high-rise apartment, later with a friend, and finally in a small basement apartment. Her activities limited by heart trouble, she relied on the government's recently increased old age pension for income and on television for company. When her son returned to Vancouver in the early 1970s, they embarked on the preparation of her memoirs, recording eighty hours of interviews and producing nine hundred pages of manuscript that were then folded into a three-hundred-page book, *A Very Ordinary Life*. It was a wonderful exercise. She concluded: 'I can't speak for others but I know I would enjoy reading this. In fact I did like most of it. It's a little too serious; there should have been more of the little humourous incidents that happened. And it's sort of strange to read about events of your own life in more or less your own words. Still, there's a world of difference between reading about something and having lived it. Not bad, pretty good in fact. But it's not the real thing.' Indeed, she and Rolf had created a remarkable document. It depicts the world of a 'modern' citizen, one who adjusted to a society and to dimensions of time and space that were quite different from those perceived by Elizabeth Goudie.

I would like to introduce one small clarification at this point. A reader might still be wondering whether it was entirely through genealogy, through the family connection between today's citizens and yesterday's founding mothers and fathers, that the people of Canada's first century served as the founders of today's country. Phyllis Knight's story illustrates why a genealogical view of the continuity of Canadian history is inadequate. She emigrated from Germany at the age of twenty-eight in the late 1920s. Her arrival, recent in terms of the nation's development, represented just one drop in the wave of six million immigrants that washed over the country between 1860 and 1940; moreover, only slightly fewer emigrants *left* Canada in the same span. That the national population increased from three million (1860), to six million (1900), and to twelve million (1946) testifies to the character of this movement: many of the newcomers were young adults aiming to found families, while many of those departing left family members behind. The pattern of population change, and of immigration to and emigration from northern North America offers little consolation to those who are looking for the essence of Canada in genealogy. As sociologist John Porter suggested, the country could be likened to a station situated at the midpoint, not the terminus, of a railway line.[2]

Commenting on the differences between Germany and Canada, Phyllis wrote: 'It's hard to put all those feelings and experiences on coming to a new country into words. On the one hand it wasn't all that strange, apart from not knowing the language. Basically, after all, we were all part of the modern world.'

II

No individual, Phyllis Knight or any other, can represent an epoch. However, the experiences and perceptions of two individuals can illustrate the dramatic changes in cultural context that have occurred in the last few centuries. The trick is to find sources that reflect many of the dimensions of a person's life and feelings. The memoirs prepared by Elizabeth Goudie and by Phyllis Knight offer the sustained intellectual and emotional effort that is required. By juxtaposing them, one can observe the scale of cultural change between the world of settlers, shaped, as we saw above, by nature and other people's texts, and the world of national citizens, shaped by the first generations of mass media, by early versions of North American capitalism, and by the boundaries of the nation-state.

The very language of the Goudie and Knight memoirs reveals differences in their authors' visions of time and space. In her childhood years in Germany, Phyllis Knight worked at household tasks just as Elizabeth Goudie was doing in Labrador. Phyllis reported that she and each of her siblings, once they had reached the age of five or six, had 'a separate task, peeling the vegetables, mopping, cleaning the house, washing. Sometimes it would be very pleasant if we were all together. We'd sing in three and sometimes four part harmony. But when the boys were a bit older they weren't expected to do any housework ... By the time I was ten or eleven I did almost all the housework, except the cooking, myself.'

Phyllis's choice of words to describe these childhood labours differed from that of Elizabeth. Where the settler depicted her Labrador upbringing as strict and disciplined – that is, she placed her emphasis on parental perceptions of a child's education – Phyllis's experience in urban Europe led her to speak a language conditioned by the market: 'Kids were thought to be the property of adults, not just in our family but by everyone.' Describing her father's printing enterprise, she said: 'My father exploited us kids something awful. For instance, he once had a big job of collating and stapling and trimming. All of us children had to work on that as soon as we came from school, for hours into the night, almost a week. And we were supposed to get some spending money for that. But instead, he bought one big chocolate bar, for all us four kids, to divide up. That was our payment.' Phyllis's selection of words to write about her childhood tasks – exploited, spending money, payment, property – distinguish her life's context from that of Elizabeth, even though both were speaking of the common experience of young people everywhere – that some of the work associated with the family fell to the children. It represents a different approach to time, one that measures it in precise chunks and that places a monetary value on it.

I am not concerned here with perceptions of personal happiness. I am addressing two different views, written in each case by individuals late in life, concerning the nature and purpose of domestic work within the context of family relationships. I do not mean to suggest that the Golm family operated only on market principles. Phyllis's memoir contains heroic stories of family solidarity, including her own risky expeditions to obtain food for the others during the First World War. However, as is evident in her choice of words, the new age had created a new language, including such notions as property and wage, that could be applied even to close and supportive relationships.

We can see the contrast between Elizabeth Goudie and Phyllis Knight in their perceptions of the relation between home and workplace. Their husbands, Jim and Ali, respectively, had to live away from home for long periods, Jim on a trap line and Ali in lumber, rail, hydro, and construction camps or in ships' galleys. The income that they earned went directly to the support of the home. However, Jim studied the land and resources of Labrador and was attuned to nature's rhythms. His work was an extension of his life; his household integrated the two units of experience – work and the rest of life – wherever the particular task might carry him. Ali's experience differed. He was restless and must have held several hundred jobs in the course of his life. Phyllis recorded one brief span when he worked in a giant bakery in Vancouver, then painted houses, operated a small bakery, travelled to Europe, returned to another small bakery, worked as a railway cook on the Lethbridge line, sailed as a ship's cook (three different assignments), then painted houses again, 'All in a year or so.' 'He was always pretty lucky when it came to jobs. Except for two years in the worst of the depression, he always managed to get a job of some sort. Although most of them were pretty lousy jobs and usually he didn't keep them very long. Ali only stayed on most of those jobs four or five months, often less, some as little as a week. He only stuck it out on a few jobs in his life for over a year. In between, he was continually running around searching and talking to people he knew about jobs.'

One obvious – but inappropriate – response to such a list is to blame the restless one and to say that, if only he had settled happily to a single task, he might have found a happier arrangement for his family and himself. Yet Phyllis respected Ali's ambition and his commitment to work. She sympathized with his choices, supported him, and recognized that there was a compelling logic behind his discontent. He would not have earned as much in a low-wage job in the city, as thousands of camp workers knew well, and the working conditions in many of the camps really did become intolerable very quickly.

What is the difference between Jim's and Ali's perceptions of household and workplace? It lies in part in the perception of the labour itself. In contrast to Jim's integration with the environment, Ali felt less involved in a more exclusively economic activity. During the war, for example, Ali worked for nearly two years as a painter in the shipyards, where 'the pay wasn't great but it was regular.'

> Ali felt like he was working his life away with no results, with no end in sight. He wanted to build up a stake so he could get away from that dreary run. But how could he, how could anybody? Then, he started to get sick from the paint. A lot of painters got that. It was a slow form of lead poisoning, I sup-

Ali and Phyllis Knight with family dog, Vancouver, c. 1939.

pose. It was especially bad when they were painting in the holds. The respirators they wore weren't much help. Towards the end he was sick almost all the time; he lost weight and he actually began to look green. So he gave up painting and went back to baking.

He worked in the big bakeries, Weston's and a few others. It was deadening assembly-line work. Ali was a master baker, that was what he had apprenticed at, and machine baking or factory work in general was something he didn't like. You did one or two operations and had to keep up with the machines. He was always dead tired when he came home, and the pay was pitiful. Well, what can you say? It was work, come home and go to work again. There's no romance in jobs. At least not those jobs. Nothing happened. People counted the various half days and small outings and the few hours they had free, maybe in total a few days over months, as great events. A day's picnic at Wigwam Inn, a boat trip to Bowen Island, maybe a movie and a meal out in a restaurant with the family. They stood out because otherwise it was work, work, work.

The jobs did not fit Ali's dreams because of the work itself and because of the pay. He was alienated from both, and, although he understood clearly the necessity of a wage, this family-driven pressure did not make the work any more attractive. Both Ali and Phyllis would have recognized the merit of economist Robert Heilbroner's suggestion that such types of income-earning activity 'desacralized' the emotional meanings of work: 'Perhaps the most striking aspect of this desacralization affects the conception of labour itself, wrenched free of its traditional social and psychological meanings and reduced to general capacities for motion and performance that enable it to be utilized exactly as if it were a versatile, but sometimes balky, machine.'[3]

The difference between Ali and Jim was made greater by their views of the household enterprise that their work sustained. Unlike the Labrador household, Phyllis and Ali's work in British Columbia pulled them in different directions, and they sometimes lost sight of common goals. The dreams and the jobs actually divided the family: 'After Kitimat he couldn't or didn't want to find any work at all for a while. Finally he took a job on the Kildala Transmission line, way back in the Coast Mountains. That was the last straw. The job was so isolated that you could only get in and out by helicopter. He wrote me a letter from that camp that really hurt me. In this letter he accused me of always forcing him to go out into the camps and that his whole life was being spent in those prisons just to get the money I frittered away. But that's not true. On the contrary, I always told him to stay in town. I didn't like him being gone all the time, away for months and months ... But no, he always went back. The good wages that he did make in those camps were just used up when he was in town without a job.' The pattern is clear: unhappy, low-paid job near home or higher wage in isolated conditions away from home. In either case the payoff was an impersonal reward – the wage – that sustained the household, more or less, at the cost of the human absences that undermined it.

Phyllis, like Elizabeth Goudie, understood from an early age the difference between girls' and boys' work. Like Elizabeth, she took on wage work when she was in her early teens to supplement the family income. And, like her Labrador contemporary, she continued to work at 'women's' tasks in the household after she was married, although she also accepted wage labour. Nevertheless, Phyllis's gendered wage work was not identical in character to Elizabeth's labour in home and field. Assumptions about gender roles complicated the experience of the Knight family.

The Labrador woman had the comfort of work that was integrated with the seasonal round, that was shared with other family members including her husband, and that contributed tangible items – fish for food, pelts for clothing – that were necessary for the daily existence of every family member. The entire family contributed to this common, work-driven enterprise. Moreover, Elizabeth's household labour could not be equated in monetary terms with her husband's activity on the trap line, because most of their work was conducted outside the market. The Goudies saw themselves as a genuine team, not as workers in unrelated enterprises.

Phyllis, in contrast, worked for wages for over thirty years: factory work, office jobs, housecleaning, caretaker at an Oddfellows' Hall, gold prospecting, bake shop manager, 'lunchman' in a timber camp ('six hundred to seven hundred sandwiches a day, about fifteen different kinds'), and cook's helper at an isolated mine where the young family was together and happy. When they moved to Vancouver she tried her hand at producing angora (rabbit) wool in her backyard – an economic failure – and she worked conscientiously in her garden. In her late forties, she joined the night shift in the sausage room of a packing plant:

There must have been about two hundred girls on that shift. We worked in a big hall, although that's not the correct term. Well, anybody who's worked in a big factory knows the sort of thing I mean. The smell was terrible. You could smell the ammonia and rotting offal from the fertilizer plant that was in the basement four floors below. Talk about pollution ...

At first I thought I'd never be able to stand it, the hard work, the night shift, the cold. It was in the winter. You had to be pretty fast, as in any factory. Slowly I got used to it, to an extent anyway. But for the first couple of months I could hardly drag myself home after work. I wasn't that young anymore and I really began to feel my age [she was then forty-seven or forty-eight] ...

I had intended to work at Burns for a couple of years if possible so that we could save up enough money to finish off the house. But I just couldn't stick it out that long. I had colds and coughs and aches and pains ... The really bad thing about it was the cold and water drafts. You had to wear knee-high rubber boots and a big rubber apron that covered your whole front, and then you still got wet. The pay was pretty poor, too ...

Nevertheless, there were some compensating factors in getting out to meet people and feeling that you had a little extra money on hand ... Most of it went into the saving for the house. I made a few new friends ... Sometimes Kay would come over to our place and have breakfast. Sometimes we would just meet Ali as he was going to work ...

Finally I got so ill that I had to quit. Always wet and cold and drafty and with your hands covered in brine or walking in and out of cold lockers. Still, there were five women ready to take the job for anyone who quit. Jobs were very scarce, especially for married women who didn't have a skill. Altogether, I worked there about eight or nine months.

Phyllis Knight's labour produced money, a sum that one could measure against the efforts of Ali and of her neighbours. She could convert it into food and shelter and clothing but didn't always obtain what the family needed. Earning a wage was not the same as producing the immediate necessities of life in Labrador but introduced abstractions into the heart of daily activity. Converting the wage into sentiments and goods demanded sustained emotional effort.

The oldest Goudie child, Horace, probably differed radically from the Knight's son, Rolf, for the same reasons that the lives of their respective parents took place within different cultural contexts. Neither child enjoyed school, it would seem, and each found his father's existence more interesting than life at home. Horace first travelled to the trap line at the age of eleven and completed less than eight grades of elementary school; Rolf sailed on a coastal freighter at fifteen, having completed most of grade eleven. Horace followed in his father's footsteps throughout his life, working in the woods, the fishery, and the fur trade. Rolf started out on this same path – the aluminum smelter project at Kitimat, PGE rail construction, the Goldbridge hydro site. Phyllis feared that 'he would get trapped in the camp life. That can happen easily; big money, big times for a few years and the rest of your life always stuck away in the bush,' as she put it. However, having had an urban experience, some contact with

Rolf, with puppy, Musketeer mine campsite
(Vancouver Island), 1941.

higher education, and more years in school than Horace, Rolf
returned to school. A succession of city jobs – taxi, factory,
construction – enabled him to get a university degree, and
even though he still worked in camps from time to time, he
turned to study and writing as well. The city and an education had given him a wider range of options.

Phyllis, Ali, and Rolf Knight grew up within a society that
differed from the world of Elizabeth Goudie. In part, the
change was a product of the job, of new stresses in the household, and of the wider educational opportunities available to
children. It arose principally from the separation of work and
wage from the production of food and shelter – from nature
itself. Northern North America had acquired new yardsticks
for a life. It had become 'modern.'

III

How have Canadian scholars explained the differences between the family experiences embodied by the Goudies and the Knights? In general, there have been two tendencies. Progressive Canadian social historians of recent decades have used other national models in depicting the world's transition from peasants to proletarians, as it is sometimes described, or from feudalism to capitalism. Those who have defended conventional views of national history, led by conservative political historians writing between the 1920s and the 1960s, have stressed more the convergence of technological and political factors in the creation of the modern Canadian state. In their view, it is the very existence of Canada as a new nation and its successful adaptation to the new global economy that distinguishes the traditional society of the fur trapper and timber cutter from the modern society built on factory, city, and consumption-based households. Both the progressive social historian and the national political historian rely heavily on economic and material forces in their interpretations of historical change. Although they have added notable chapters to the story of Canada, neither, I suggest, offers an approach that takes into account the dramatic differences in today's readership from that of yesterday's. Neither type of account presents a satisfying explanation of how the common people have shaped the development of Canadian institutions. Both are too narrowly economic, and insufficiently cultural, to convey the distinctiveness of the Canadian experience.

The pivotal new force in the lives of the Canadian people, according to progressive social historians, was the Industrial Revolution. The famous technological developments that refashioned the British economy from the mid-eighteenth century on, they suggest, constituted not just improvements in machinery but a revolution in human relations. They point out that common people had to learn to obey market forces when they sold their labour in exchange for the necessities of existence. They emphasize that common people had to deal with new types of work when they encountered giant factories, new forms of society when they moved to burgeoning cities, new skills for daily communication when they picked up daily newspapers. According to these social historians, northern North American society, like every society around the world, was being reshaped by capitalist assumptions.

These social historians contrast the new society, the one in which the Knight family made its way, with that known by Elizabeth Goudie in Labrador. They describe Elizabeth's world as one of family-based production and emphasize that it relied on necessity and parental discipline as spurs to action. In times of crisis, they note, Elizabeth and her contemporaries provided reciprocal assistance among families in the form of gifts of shared food and hardware. This reciprocity was based on the community-wide assumption that everyone had public, or God-given, rights to resources. Economic activity in the settler societies, the new social history contends, could never be separated from people's religious, political, and social beliefs in the way that such distinctions came to prevail in a capitalist environment.

By placing the Industrial Revolution at the centre of their vision of world history, the new social historians are proposing that its capitalist-based production reshaped both work habits and domestic life. They say that the market, in exchanges conducted under capitalist assumptions, reduces personal sentiments of sympathy and obligation to the logic of monetary calculations. No other productive system, they write, has been so successful in creating a single worldwide language of comparison for all types of human activity – the language of money. Capitalism asserts, they claim, the appropriateness and even creativity of amassing wealth, or capital, and using this surplus to create yet more wealth. They cite this as a fundamental distinction between capitalist-based and other societies: capitalist systems use wealth not simply as a material thing – like a machine or a painting or a landed estate – but as a *process*, a continuously renewed, impersonal process of profit-making in which the profits must be reinvested and put at risk. The profits then transform themselves into further surplus or move to competitors.[4]

If market and capital constituted the 'heart' of the new order, in the thinking of these social historians, then three additional processes – property, wage and commodification – were its life-blood. Property is a bundle of rights, in this view, including the right to use and to prohibit others from using land or tools or capital. Today's property system evolved over many centuries, and its basic principles hardened into unassailable law in northern North America only in the nineteenth century.[5] Its introduction, especially in areas of Aboriginal resource use, occasioned many misunderstandings. In former times, the product of one's labour belonged to the person who did the work, as in the case of a farmer's grain or a hunter's furs or a woodcutter's logs. However, when produced by wage labourers in the capitalist order, such goods were allocated to investors who might do none of the physical work. One's rights to one's own product, as the farm or fishing family might have described such property relations, were thus redefined to safeguard the flexibility and liberty and creative capacity of the entrepreneur. The change might come slowly, perhaps because of shifts in investment practices: thus shares in the harvest of the Great Lakes fishery were allocated originally to both the fisher on the boat and the investor in the portside office. However, as the years passed and the investment in the business increased beyond the levels available to an ordinary labourer, the investor assumed control, including absolute ownership of the fish that had been landed by someone else's labour.[6] Working people's relation to the natural world, according to this interpretation, changed irrevocably with the revisions in society's approach to property: henceforth many ordinary citizens, particularly those who worked for wages, related to the land through another's ownership of place and tool and product.

Social historians see the worker's wage not just as another type of income, identical in character to money earned by the sale of wheat or furs, but as a new form of human relationship. Because of their position as buyers of labour through the payment of wages, the owners of enterprises possessed property in the form of product and tool, but they also commanded the company's operations and disciplined its workers. Thus the wage carried fundamental assumptions about power and freedom. Employers might well believe that they alone should determine the working conditions and choose the location, materials, pace, and even purpose of work.[7] Of course, the workers might disagree with this view. Their struggle itself redefined time, which became the span of work for which a wage was paid. It also restructured space, because it separated workplace from household, as the differences between the domestic circumstances of the Goudie and of the Knight families illustrated.

Social historians have emphasized, too, the capitalist system's drive to innovate. This economic order, above all others, rested on commodification. Greater efficiency and newer, more satisfying products were always imaginable. Thus the system rewarded the identification of new human needs and the development of new production techniques. Whether dealing with a physical good such as newsprint, or a service such as advertising, the innovation would, if it produced a return on investment, find an entrepreneur eager to commodify another corner of life. Like wage and property rules, commodification transformed the environment in which one earned a living, according to the new social history. It also required rapid and continuous adaptation on the part of ordinary citizens.

Such changes constituted an extraordinary force for uniformity. They also made the experience of ordinary citizens in Canada resemble those of ordinary citizens in every other industrializing country in the world. What was distinctive about the Canadian story? Here the new social history ran into problems. It assumed that there was a Canadian cast to this transition, but it spent much of its energy demonstrating the similarity between Canadian experience and that of a vaguely defined international norm. Naturally, the world leaders in the adaptation to industrial capitalism, Britain after 1750 and the United States in the twentieth century, often provided the implicit comparative context. And, not surprisingly, Canada often came out looking distinctly inferior – not enough class conflict, too few movements of resistance. This is hardly a story to inspire youthful patriots. It may be inclusive and socially aware. It certainly distinguishes the modern epoch from the traditional societies that preceded it. But is it all that can be provided for today's generation as a history of northern North America?

Many national political historians have responded more positively to the introduction of capitalist principles in northern North America. They believe that the century after 1840 was distinguished by nation-building and economic growth. Thus they emphasize the change from small, resource-based colonies in the mid-nineteenth century to a relatively productive nation exercising considerable international influence in the mid-twentieth century. Where the social historians count the social costs of the Industrial Revolution, the national political historians prefer to enumerate the sources of wealth and to outline the partisan struggles for power. By implication, they attribute Canada's favoured status among nations in the 1950s and afterwards to the cultural advantages of a northern, Britannic, and capitalist community.

National political historians accept without much debate the idea that the rapid increase in northern North America's wealth production resulted from its development of a market-based economy. They emphasize the diversity of this activity – in four major areas – and laud the country's rapid economic transition. First, farm-produced wealth provided the foundation for Canada's growth, even though agriculture's relative role within the national economy declined steadily.[8] Second, the new abundance, they say, also derived from resources located in forest and river and precambrian shield. The original staples in Canadian history were fur, fish, timber, and wheat; the new staples included the four forest-based successors to squared timber – lumber, wood-pulp, paper, and newsprint – as well as hydroelectric power and a long list of minerals.[9] Third, manufacturing, the engine of Britain's Industrial Revolution, accounted for more than one-fifth of Canada's gross domestic product from the 1870s to the 1980s. This may seem a small proportion of the total, but national political historians see the emergence of major enterprises in this sector as signalling Canada's membership in the new economic order. Canada did not become one of the greatest manufacturing nations, but it was closer to the European and American pattern than to the economies of Latin America.[10] Fourth, delivery not of goods but of services became essential – marketing, finance, education, health maintenance, and many others. The service sector grew steadily, from 30 per cent of productive activity in 1870 to over 60 per cent by 1945.[11] Partly a statistical phenomenon caused by the shift of formerly in-house activities from, for example, manufacturing or mining companies to securities and advertising firms, the increase also reflected the com-modification of modern life. Canadians, like their contempo-raries in other capitalist economies, increasingly relied on the market for the supply and pricing of goods and services. Thus, according to this school of historians, Canada travelled on a global economic tide but benefited from a favourable combination of resources and people.

Each person's economic production (gross national prod-uct, or GNP, per capita) increased by a factor of five between 1870 and 1950. This index, these historians imply, offers one of the central lessons in Canadian history: *each* Canadian in 1950 on average produced five times more wealth – as mea-sured in money terms – than his or her 1870 counterpart. Using constant 1981 dollars to establish a meaningful gauge, they observe that the country's economy produced just under $5 billion in goods and services in 1870 and just over $83 bil-

lion in 1950. When compared to the era of family-based subsistence and commodity production, they alleged, the
economy was growing faster after Canada's industrial revolution and enabling each worker to be far more productive.[12]
Moreover, the change was a coast-to-coast phenomenon.[13]

These national political historians face a problem that is a
mirror image of the one faced by the social historians. Both
are crafting their stories from economic materials. Where the
social historians see conflict and unequal distribution of
wealth and power, their national political rivals prefer to
emphasize growth and the prospect of prosperity for all. Yet
just as the social historians find great similarities between
Canada and other modernizing nations, so too do the political historians. And here's the issue: if all these countries are
travelling the same path, what distinguishes Canada from
the rest? Given that the United States led the world into its
twentieth-century version of industrial capitalism, is Canada
not simply slower to adapt or, perhaps worse, an inferior
imitation? In either case, the assumptions of the national
political historians do not adequately explain Canadian distinctiveness during this first century of national development and do not inspire visions of Canadian creativity in an
earlier age.

IV

Farm, class, and gender can illustrate the debate between the
new social and the national political historians. Though just
three among many possible examples, these subjects suggest
that it is often the angle of vision, not the topic, that distinguishes the two schools.

National political historians celebrate the expansion of
Canadian agricultural production in the late nineteenth century and the emergence of a wheat-exporting economy on
the prairies in the early twentieth. The new social historians acknowledge this rural vitality but tell a different story:
the farm family did not differ substantially from the wage-

earning household in its lack of power and relative poverty.[14] Thus these social historians see the prairie farm as possessing an ambivalent status. Certainly, it did not obey all the rules of modern capitalism. It differed substantially, for example, from the estates in late-nineteenth-century Britain and northern Europe, where employment of wage labourers was the norm. Because of the economic advantages of household production, the prairie farm was in fact superior as a producer of grain to such large, estate-run, wage-based enterprises. The farming family was flexible in its purchases and in the amount of work required of its family members; moreover, it was prepared to postpone profit on its investment for years, as long as the enterprise itself remained afloat.[15] This view permits social historians to argue that many decades passed before the conventions of the capitalist nation were consolidated. Thus they can claim that Canadians struggled over these rules and in the process defined their own distinctive society. There is much merit in this view, but, as I noted above, the process is similar in most industrializing countries.

National political historians prefer to concentrate on the eventual victory of the new economic order. They believe that, although the family farm's labour system seemed family-based, capitalist dimensions increasingly prevailed in much of farm life, as they did around the world.[16] They suggggest that members of these farm families increasingly worked within fluid, impersonal markets not just for grain but also for land, credit, implements, groceries and even, in moments of pressing need, labour. Again, there is merit in this argument. It is true that the farming household purchased a greater proportion of its inputs in 1950 than in 1930, and more in 1910 than in 1890. An inexorable process was pulling it farther into the market-place. The trend was evident in the equipment shed and the kitchen but also in the mental calculations of the people. Increasingly, family members talked not of an eternal stake in the land but of annual profit and loss. They compared the farm's return to that available in other employment; they estimated the contributions of animals and tools, not for a lifetime, but in a seasonal cost-benefit equation that set horse against truck and tractor; even home preserves were put to the test of the marketplace; and the new language of rural sociology spoke of investments – of putting dollars in the rink or donating time to the youth group – as the price of a viable community. In the view of the national political historian, then, the farm family treated life on the land as just one economic option among many. A wage-paying job offered merely an economic alternative. Prairie farm children learned to weigh farm and city career options in the universal currency of dollars and cents, the language of the market.[17]

This debate has produced valuable new understandings. However, an unheroic middle stance offers as much illumination as either pole. The prairie farm household of 1925, for example, occupied an intermediate position within industrial capitalism: it was not peasant, as in the case of medieval serfs; it was not a settler household, in the fashion of the Goudie family of Labrador or the Métis buffalo hunters of the prairies; it was not proletarian in the way that Phyllis and Ali Knight of British Columbia and their wage-earning contemporaries in Canadian factories and work camps were. However, these farming and wood-cutting rural residents did operate in an increasingly capitalist world. The conventions of the larger society, including the fundamental dimensions of time and space, were increasingly national and based on literacy. In such a context, how can one dispute the assertion that rural families were shaped by a print-capitalist environment? Although the new social historians may emphasize the virtue of resistance to the new order, they must recognize that the national political historians are on the side of the angels in this discussion: after all, capitalism is the dominant economic system in the world today.

The new social historians placed gender and class at the forefront of their studies of Canada's first century. They recognized that working people endured several generations of poverty as the industrial revolution transformed Canadian economy and society. They also perceived that this extraordinary social and economic event differed in its impact on men and on women. National political historians have been slower to take up the challenge posed by this significant reorientation of historical study.

Gender, which can be defined as knowledge about and social organization of sexual difference, is often assumed to be a given. However, it differs markedly from country to country and class to class. As Phyllis Knight's story illustrates, making a living in a blue-collar household required that the wage labourer work within a factory or mill or camp, while the provision of food, clothing, and shelter, not to mention the raising of children, remained under the family roof. The family norm in Canada was a single breadwinner: 'no two,' said a workman in Hanover, Ontario, to historian Joy Parr, in explaining his assumption about one breadwinner per household; 'there's got to be somebody home with the family.' The rule did not prevail simply because of a male's place in the wage-earning market; rather, as Parr explains, the males' authority depended on 'the social definition of their familial position.' What does this mean? Although the breadwinner in the wage-earning, out-of-home role was not always male, and even though the leading breadwinner might provide less than the entire family income, the position of senior male in a household carried with it power far beyond the man's relative economic contribution to the family unit. Whether owners or workers, these men held in common an assumption that male dominance *should* constitute 'the fit social distribution of resources and power.'[18] Within the narrow range of options that he had, Ali decided about his own work, and Phyllis had to manage in the economic context thereby created.

For most of these years, the material condition of families reliant on blue-collar wage-earners was precarious. For example, not before 1928 was it possible for the average male manufacturing worker in Montreal to maintain a small family on his wage alone.[19] Of necessity, other household earnings, often women's, whether from piecework, from keeping boarders and maintaining garden plots and small numbers of animals, or from wage labour, supplemented the family income. As soon as was feasible, children also went out to work, earning the small sums that might balance the household's budget. The fact that women as well as men subscribed to the 'male first' and breadwinner approaches ensured that men would receive twice the female wage and inevitably that gender biases would normally be reinforced by task specialization: men outside the home, women within.[20] Men might contribute some of the labour associated with the garden and much of the structural repair, but women managed domestic arrangements (keeping expenses lower than revenue); bore and cared for children (including passing on the crucial elements of the family's heritage); looked to the well-being of the working adults, the care of dependent adults, and the housework; and, if market avoidance and barter did not suffice, earned sufficient money to keep the household from foundering. Because the household often required additional labour, and because the generally accepted notion of a 'family wage' undercut women's wage rates, daughters were more likely to work at home, sons away from home.[21]

National political historians concentrated on party, region, and nation when they dealt with Canada's first century.[22] They said that because Canadians never experienced the intense conflict that marked Europe's industrialization, class did not belong in Canadian historical discussions. Like the rest of the historical community in the pre-1970 generation, they rarely considered women's place in national history at all. They recognized differences among residential districts, in worker–supervisor relations on the shop floor, between the consumption patterns or club memberships of the wealthy and the poor. However, many rejected the view that these differences might be enduring, and many doubted that differences of language and religion and region would ever be dissolved in a pan-Canadian class solidarity.[23]

I suggest that, beneath these debates about the past – which were also, as is always the case in historical writing, debates about the politics of today – the two schools had much in common. This resemblance was a consequence of the inescapable truth that all Canada, along with the rest of the world, was undergoing an economic transformation. Manufacturing illustrated the pressure to conform. American ownership and management of Canadian manufacturing plants increased with the imposition of the National Policy tariffs in 1879 and became a formidable force in the twentieth century with the rise of new mass production industries, such as automobiles and electrical goods. Most of these branches were built adjacent to their American parents, in the booming manufacturing belt of southern Ontario and Quebec. The U.S. share was more than half of total foreign investment in Canada by 1926. By the 1930s American capital controlled nearly one-quarter of Canadian manufacturing, including most automobile production, two-thirds of all electrical goods, and about half the newsprint industry.[24] U.S. ownership usually entailed American management methods, including the 'scientific' approach to organizing tasks, improving efficiency, and establishing a hierarchy of responsibilities, all determinants of workplace atmosphere.

Many Canadians lived in a single-industry town that relied on a hydro plant, a paper company, a mine, or a railway for its lifeblood. Some, like Ali Knight, worked in remote, men-only camps. Yet despite the variation, the underlying economic institutions were the same. Canadian capitalism operated within an imperial and North American – for which one should understand 'absentee' – system of ownership. Although the British North America Act of 1867 placed land and its resources under the jurisdiction of the provinces, and the new staples did not act as nation-building economic activities in the way that prairie wheat did, the crucial instru-

ments of power rested in distant metropolises, where financial, technical, and purchasing enterprises had their corporate headquarters. Thus, north-south trade, American
influence, and province-building constituted major themes in
Canadian resource development. Ironically, they exacerbated
both regional economic difference – Ontario paper, Alberta
coal – and continental and transatlantic capitalist homogene-
ity. Only the *politics* of the provinces became significantly
more 'regional' (that is, fractional or anti-central); the econ-
omy-wide trends fostered similar pressures in every North
Atlantic workplace.[25] Thus the social and the national political historians were dealing with a common phenomenon.

   To the two schools of history, what mattered was whether
one looked at these momentous changes from above or from
below. In the view of social historians, people like Phyllis
and Ali Knight constructed their lives on wage-based relations with managers and bosses and within class-segregated
neighbourhoods. They had little control over the powerful
constraints that shaped their daily existence. National political historians, who were more likely to address the views of
the owners, suggested that the great corporations assumed
leadership of most Canadian communities during this century. The maintenance of order and the conduct of smooth
productive business relations were the concerns that figured
prominently in their work. Both types of historians would
have agreed that, because of Canada's federal system and its
scattered, fractious workers' communities, ordinary citizens
were slow to establish campaigns aimed at reforming public
policy and ending instability.[26] As E.P. Thompson said of
English workers in the late eighteenth and early nineteenth
century, the first generation within these new capitalist
dimensions had to be taught by their masters the importance
of time before the second generation could fight for shorter
hours and the third generation could insist on extra pay for
overtime.[27]

   The two historical schools also disagreed about the middle
class. As the incomes of certain occupational groups rose

(especially relative to others), as new products became available, and as tastes changed, consumption patterns also came to distinguish social groups. Clothing in particular declared an individual's wealth and aspirations. Segregation by neighbourhood and by membership in recreation and service clubs reinforced the growing sense of distinctive group identity. Most of all, although this is hardest to discover, family strategy probably marked some people off from others: for example, some people delayed marriages, had fewer children, exercised greater supervision over their children's formal *and* informal education, encouraged longer schooling for them, and displayed extreme caution in the so-called major life decisions. Such characteristics supposedly distinguished many aspiring white-collar families from their blue-collar contemporaries.[28] Did these people adhere to one group identity? Did their interests and their ideals conform to those of workers, to a separate reform agenda, or to the larger capitalist project of the age? One could make a case for each of these views. One might also argue, however, that all these options belonged within a very narrow range. The debates between social and national political historians relied on an economic approach to history. Neither version provides the distinctive Canadian theme that might inspire interest in the national narrative among today's citizens.

V

I am suggesting that the social and political approaches to the national story reflect preoccupations that are not as relevant today as they once were. Although political battles over the nature of this capitalist order are still being fought, the historical battles over whether to celebrate capitalism's creators or to take the side of its opponents seem out of date. Instead both sides are struggling over yesterday's news. The cauldron that was capitalism has melted everyone into a new and, from the perspective of a nineteenth century settler household, relatively uniform society.

Why the uniformity of experience? The answer lies partly in the capitalist institutions themselves – market and wage and commodification and private property – but also in the transportation vehicles and communication media on which they travelled. As the residents of northern North America built railways and cities and factories, they joined other industrializing countries in believing that they were making the leap from relative scarcity to relative abundance and from local isolation to international interconnectedness. Distances might be overcome, the nation made viable, only by a revolution in communication technology. Several innovations, including the telegraph, the newspaper, and the railway, permitted political leaders at the time of Confederation to imagine the survival of such a Canadian empire.

In the nineteenth century, the railway seemed to diminish physical distances and bridged previously insuperable obstacles to travel. As Maurice Careless suggests in his discussion of the United Province of Canada in the 1850s: '[The railway] quickened the very tempo of Canadian life. No longer need it move at the pace of the lumbering ox-cart, or even the thrashing steamboat, when the broad-stacked locomotive, belching wood smoke, clattered the cars at speeds up to forty miles an hour across the province, and its whistle screeched "down brakes" at once isolated inland hamlets. The railway, besides, lifted the old, retarding barrier of the winter freeze-up. Farmers and shippers saw goods flow regularly and speedily to distant markets ... Newspapers printed in the cities circulated ever more widely; ... and railway progress brought a new degree of opulence that was vaunted in "elegant" public and private buildings still scattered through the towns of Ontario and Quebec.'[29] The characteristics of space in Canada had changed because the country had grown smaller in effective size while it had expanded in actual territory.

If rail lines tied physical spaces together, the telegraph that accompanied them effected a remarkable divorce between transportation and communication. What was once synonymous – the movement of physical goods and the dissemination of information about them – separated decisively when electronic telegraphy made possible the rapid transmission of data. Markets were revolutionized because local conditions of supply and demand no longer dictated price. The railway and the telegraph could combine to transcend local dearth or surplus by moving goods reliably to places where they were wanted. Buyers and sellers everywhere could now assess their options accurately and respond quickly to the changing conditions of supply, demand, and price.

By creating a 'level playing field' for markets across space, this higher quality of information permitted development of markets in futures. Thus, where merchants had once speculated on price differences between locations, they could now speculate on differences in time (for example, the production of wheat in Kansas next June versus that in Manitoba next August and in New South Wales or the Argentinian pampas next November), adding an entirely new dimension of uncertainty. Commodities were thereby 'moved out of space and into time.' The markets in which they were sold became more abstract, and the good itself – a bushel of wheat, for instance – lost its physical uniqueness and became interchangeable with every other one of like grade. It could be reduced to a mere likeness, such as a warehouse receipt. As Karl Marx recognized at the time by introducing the idea of 'commodity fetish,' trading in futures transformed a mere product into a more complicated entity, a commodity. The changes in communication 'had the practical effect of diminishing space as a differentiating criterion in human affairs,' according to James Carey, the scholar of communication. The price system ensured that everyone lived in the same place or space because markets were 'decontextualized.'[30]

In sum, capitalism and communication technology restructured the economic consequences of space and time in northern North America. They introduced new dimensions that were shaped by market, wage, property, and money, as well as by the media of rail and telegraph. This power to reorganize society is evident in the different experiences of the Knight family in British Columbia in the 1930s and those of the Goudie household in Labrador in the same decade. The land and its resources – including the food and clothing and shelter that one could produce from those resources, which the Goudies knew so well – now were constrained by boundaries and ownership rights that the Knights had to accept. The goods on which one built a life were henceforth to be measured and priced by national and international standards established in competitions with similar goods from other parts of the globe. The global capitalist process in turn attached new relevance to the costs of location and transportation. Even household space acquired new, gendered qualities as production and reproduction, urban and rural, white and blue collar, acquired new implications. Thus Phyllis and Ali Knight measured their lives against different standards than the Goudie family, even though the milestones – marriage, birth, changes of residence – seemed identical.

How did the ordinary citizens who had to cope with the emergence of this new way of life in northern North America contribute to the growth of today's Canadian institutions? Progressive social historians provided a great service to Canadians by making evident the central role of capitalism and new communication technology in the remaking of society. National political historians used a different approach to economic history to make a different point: they wished to establish the distinctiveness of the Canadian nation within the British and then the American imperial orbit and, therefore focused more closely on the ins and outs of parties and personalities. They were implicitly celebrating the country's rapid economic take-off as a modern industrial power that distinguished it from less-prosperous, less-diversified nations. Today, both approaches, which rely heavily on conflicting views of economic history, inspire little enthusiasm. It is time to introduce cultural perspectives to the Canadian national story to see what Phyllis Knight did contribute to her adopted land.

6

Literate Communication and Political Resistance

Perceptions of time and space changed dramatically between the 1840s and the 1940s, with noteworthy cultural implications. Thus there emerged a single standard of public time. Its precise measurement became inescapable, as life with clocks, timetables, and assembly lines demonstrated. And yet the new age also introduced infinite numbers of perceptions of private time. The paradox moved some observers to conclude that time itself was merely a social construction, as variable as the uses to which it was put. As for space, it seemed to shrink. Ordinary citizens found it impossible to resist the new pace of life that the closer quarters made possible.[1] These were worldwide changes, and they introduced great pressures for uniformity. Surely all Canadians were growing more alike and more like citizens in every other rapidly developing community.

Changes in communication technology and economic arrangements transformed every sphere of life. For the previous five thousand years, those who could read and write had been, as Benedict Anderson described them, 'tiny literate reefs on top of vast illiterate oceans.' The convergence of the capitalist ethos and print technology in the sixteenth century, however, set the stage for 'a new form of imagined community,' the modern, literacy-based nation. For several centuries this new kind of community recorded the hopes and organized the political activity chiefly of elites. During this

period, 'unified fields of exchange and communication' and
the modern forms of individual print-languages emerged in
parts of Europe and the Americas. Certain 'languages of
power' dominated international conversations because of
their use in bureaucracies, the market-place, and the academy, while others (Cree and Gaelic, for example) survived
only in local communities. As it happened, the construction
of modern Canada took place in two languages, in two
administrative bureaucracies spawned by two 'Europe-cen-
tred, world-imperial states,' and in two imagined communities – French and English.[2]

More crucial than the emergence of two official languages
in Canada, however, was the sheer power of the print-capitalist mode of communication. Just as the seventeenth-cen-
tury captain of the canoe brigade followed the waterways in
establishing the flow of trade, so nineteenth-century bureau-
crats hastened the adoption of laws and regulations by circu-
lating them with the aid of library, newspaper, and telegram.
Paper and telegraph wire, it soon became apparent, had
replaced rivers and sailing vessels as the dominant media of
communication in northern North America. The nation of
Canada owes its birth to them. And with them, ordinary citizens fashioned a response to the conditions in which they
lived that has been fundamental to the growth of Canadian
institutions. Thus, the redefinition of space and time in this
epoch was accompanied by the reconstruction of the conditions of daily life through ordinary citizens' political action.
Accordingly, in this chapter I look at four crucial contexts for
this process – the national state, the media (especially newspapers), the struggle between capitalism and socialism, and
popular politics – with Phyllis Knight's story as a reference
point.

I

What is the meaning of 'nation'? What is the definition of
'the state'? Phyllis Knight absorbed the implications of

nation and government from her early days. In school, she studied a version of German history. As a teenager, she experienced the crisis of the First World War, when both her father and brother were drafted into the military forces. She carried a ration card, endured food shortages, and even broke into storage depots to help feed her family: 'In fact, we were fairly lucky. None of us were killed, and that was a time when whole families died out, when whole regions were destroyed and a whole generation was shot to pieces.' These extraordinary experiences, driven by nations and their governments, made the nation-state a matter of second nature to Phyllis as she entered her adult years.

Phyllis and Ali had dreamed of establishing a small chicken farm near Berlin that would somehow marry the architectural ideals of Walter Gropius and the Bauhaus group to the 'back to the land' sentiments prevalent in their Wandervoegel (Wandering Bird) youth movement. On their dearly won Pomeranian farmsite they discovered instead rural poverty and just another form of endless labour. Two years were sufficient to snuff out that dream. After debating the alternatives – Ali talked of South America, new lands, open frontiers, building a home in the wilderness – they 'compromised on emigrating to Canada.'

Immigration reinforced Phyllis's sense of the national and governmental dimensions of a modern life. Whereas Elizabeth Goudie never seemed to have considered a political unit of any sort before 1940, referred to Labrador in terms of the green hills, bluewater lakes, and Atlantic bays over which she travelled, and mentioned government only in the context of a welfare payment, Phyllis spoke often of nation, politics, and the policy decisions of the state. Even her choice of Canada had a political ring to it: 'We didn't expect to get rich and we didn't think the streets were lined with gold. But we figured one could lead a pretty good life in Canada, if you worked and after you got settled in ... Moreover, we half expected that you would be less surrounded by restrictions in Canada. I guess it was the feeling of getting a new start, corny as that may sound. I think what attracted us most was that Canada was still a big country with a lot of open space and nature left but that at the same time it was civilized and orderly in a way South America or even the United States wasn't.'[3]

Phyllis started out with assumptions about government, law, and control that would have been alien to Elizabeth Goudie. Thus her amazement about Canada's pattern of freedom and authority sprang from reactions honed in a distinctively modern environment: 'What we found the most strange was the pretty free and easy attitude to what you could get away with in some things and the tremendous provincialness in other areas. For example, Ali was once stopped by the police on a Toronto beach for wearing a topless swim suit. Ali, mind you, not me. That was in the first year we were there and we couldn't believe it. On the other hand, in a lot of places you could just go out into the woods and make yourself a cabin and squat there. That was just unbelievable. There was another feeling we had too. The feeling that nobody, not the government nor any group nor even most people you knew seemed to care very much about what happened to you.' Phyllis and Ali were conscious of government, even its absence.

Language occupies a special place in discussion of nation, partly because of its role in establishing the boundaries of communication but also because it channels the currents of thought in the community. Phyllis treated the difficulties of learning a new tongue, English, in an off-hand manner. However, as anyone who has acquired a language later in life can attest, this was not as easy as it sounded. Her description of the actual process illustrates the pains she took:

> I wanted a job too but I still couldn't speak any English, apart from a few stock phrases. I had a grammar book and a phonetic dictionary and I started out reading 'True Romances' and 'True Detective,' those sorts of magazines. They were just what

I needed. The stories and the words and the constructions they use in them are very simple. And the story lines and plots are primitive and predictable. I'd write down every word in English I didn't know, which in the beginning was most of them. Then I'd write the German equivalent and try to pronounce the English word. Later on, when I had a fair stock of words, I'd make up imaginary dialogues. Ali helped me but what I really needed was somebody who could really speak the language, because often my sentences weren't understandable. I made up my own phrase-book of sentences that I thought I should have for going shopping and for chance conversations in the park and things like that. The first talking movies were coming in about then and I went to them fairly often too. They were quite helpful.

All in all, I learned English relatively quickly after I had the basics down well enough so that I could carry out the rudiments of a conversation. In less than two years I read what were then modern novels in English without any trouble. Of course I used the dictionary quite a bit. In fact I still do now, when I'm reading, because there are a lot of words in print that you never hear spoken.

This painstaking study presupposed that English would be her language of communication and that she would inhabit a community wherein public discussions and works of fiction alike were only in English. Of course, this is not true, yet Phyllis never mentioned French or the place of French in Canadian life. What is more, she never returned to the topic of language itself in her memoir. In this absence, one can see the difficulties of a country that is built on two official languages and on immigrants who settled thousands of kilometres away from the largest concentrations of French-speaking Canadians. Phyllis Knight's experience is an example of the increasing power of the new communication technologies to *separate* people one from another. Phyllis in Vancouver, like Elizabeth in Labrador, had dealt in several languages in daily life. Unlike Elizabeth, however, Phyllis was enclosed in an environment wherein language itself, though invisible, erected barriers against others. Her new society was surrounded by a perimeter wall that separated those who could deal easily with the messages of print, radio, and film from others who could not. Whereas Elizabeth Goudie had no trouble dealing with Aboriginal visitors in her Labrador household, spoke not at all of the language barrier, and enjoyed the challenge of communicating with them, Phyllis felt it necessary to describe the precise measures she adopted to learn a large body of words and grammatically correct constructions. She then became immersed in the enclosed community that the language depicted – but also created. Language conveyed by the media also helped to redefine her perception of space.

Phyllis knew the difference between cultures created within the governmental structures of Germany and Canada, but she also could see the contrasts between the two communities as nations. Her name change from Krommknecht to Knight received not a word of explanation in her memoir, but several other incidents suggested the national dimension of the age. She and Ali started out in Toronto, a large city teeming with immigrants, whose mutual unfamiliarity with the host society provided them with a degree of comfort. But Toronto also contained vigorous national prejudices. Phyllis responded to an advertisement for a housemaid and was met at the door by a woman who said: '"Oh, you're German. I wouldn't dream of hiring a German girl. I don't forget the war that easily." As if I had anything to do with the war. I was sort of taken aback and my English wasn't very good when I got flustered. But Agnes, who was with me, got her Scotch dander up and really told that woman off. All that took place on the doorstep if you can imagine.'

Phyllis's experience of the Second World War offered her further reminders of national identification. Having somehow failed to be nationalized as a Canadian with Ali, she had returned briefly to Germany in 1937 to care for her mother in the aftermath of a sister's death and the imprisonment of three brothers for political activity. On her return to Canada, and after the nation's declaration of war in September 1939, she was classified as an enemy alien. She avoided registration for three years: 'I don't care what the law says, I'm not going to ... put myself in the hands of some officialdom which is full of hopped-up patriots.' When officials 'caught up with' her, a sergeant in the Royal Canadian Mounted Police threatened her with jail: 'Well, I usually didn't cry, but that time I did – from anger and frustration. Because here were the characters who could ruin your life and they knew nothing about anything. Then I said, "That's okay. You put me in jail, I don't care. You just give me a lot of books to read and leave my child with me and I'll be alright. I don't care." They had jail cells right in that building. He showed them to me to frighten me, a silly way, I thought.' Finally, the officer approved her pass, on condition that she report for an interview once a month. She described the process as 'humiliating and a little frightening,' but not more than that: 'Actually, reporting in wasn't any real burden. It was just that feeling you were under suspicion and that the government might change its policy after all and pack you off somewhere. I realize that compared to what was happening in Germany and some other countries it was a fairly mild injustice. Nevertheless, that didn't make it seem much better.'

Her observations on the resettlement of her Japanese-Canadian neighbours were more acid and again revealed her sense of nation and government: 'What a rotten bunch. The newspapers, the city politicians, the Canadian and BC governments – most of the leading lights of the country. Of all the people who had any position at all it was only a few of the CCF leaders like the two Winches and the MacInnises and some others who stood up for the ... Japanese-Canadians ... In general there was a feeling of pure hatred ... It was pretty frightening and we didn't know if we would be next. It showed you clearly just how much you could depend upon justice, if the chips were down.'[4]

The experience of the war shook the Knights' confidence in

the fraternity of the working class. As Rolf, their son, commented: 'Probably the greatest cultural shock that my parents had was the almost overnight conversion of the
workingclass left in Canada into a vociferous' chauvinism
during the Second World War. It underlined the fragility of
socialist alliances. I think it, more than anything else before
or after, left my parents feeling isolated and alienated.'[5]

Phyllis's experiences with nation, language, and government illustrated the increasing power of such forces in modern life. She exchanged German for English, and Germany
for Canada, but even then, as she observed from the experience of her Japanese-Canadian neighbours, she could not be
certain that her new citizenship would go unquestioned forever. The space in which she dwelt now was delimited by
language and steeped in narrowly defined national loyalties.

II

Phyllis loved to read. She started before her formal schooling
by copying her older brother's schoolwork: 'By the time I
was seven I was an avid reader ... What did I have to do if I
wasn't reading anyway? Either sit around or help with more
work. I always had a book with me, at home, in school, and
wherever I had to wait for five minutes, I read.' She received
a basic schooling of eight years at the local Volkschule but,
like all the children in her circle, could not aspire to higher
education: 'Almost none, in fact none, of the children I grew
up with ever went to the Gymnasium [the academically ori-
ented system of high schools]. That took more money than
ordinary people could afford ... Even most people who went
to Gymnasium never got into University. That was only for
the sons of professionals and the rich.' Nevertheless, when
she was working in Berlin in the 1920s, she did attend some
lectures and night classes. This love of reading never left her,
and later in life she took great pleasure in talking with
friends about books.

Phyllis's experience showed that nation and language are

firmly tied to literacy and print. As rail lines and telegraphs linked physical spaces, so news columns united imaginations. When Canadians took a bench in the streetcar or a chair at the table to read the *Globe* (Toronto) or *La Presse* (Montreal), they bridged both space and time in their musings. They travelled to legislatures and courts and concert halls and factory floors and hockey rinks, because that was where the news was being made. But they also could catch glimpses of other readers and reading places in the land, because there too Canadians were simultaneously reading similar or identical stories and recognizing the synchrony. As Benedict Anderson explains:

> The significance of this mass ceremony – Hegel observed that newspapers serve modern man as a substitute for morning prayers – is paradoxical. It is performed in silent privacy, in the lair of the skull. Yet each communicant is well aware that the ceremony he performs is being replicated simultaneously by thousands (or millions) of others of whose existence he is confident, yet of whose identity he has not the slightest notion. Furthermore, this ceremony is incessantly repeated at daily or half-daily intervals throughout the calendar. What more vivid figure for the secular, historically clocked, imagined community can be envisioned? ...[The ceremony required that communities discover] a new way of linking fraternity, power and time meaningfully together ... Print-capitalism ... made it possible for rapidly growing numbers of people to think about themselves, and to relate themselves to others, in profoundly new ways.[6]

Canadians' perception of space and time had changed. Now they could study maps of a vast country, engravings of distant scenes, and pictures of great parliamentary moments that they might never see in 'real life,' and they could realize that they encountered them simultaneously with their compatriots from sea to sea. They possessed these phenomena together. The dimensions and boundaries of their imaginations – very different from physical dimensions, but as real in their way as the relative distances covered by horse and railway – had changed forever.

Newspapers helped to shape people's awareness of boundaries.[7] When Confederation came into effect on 1 July 1867, and when this new state absorbed the vast western and northern reaches of the continent in 1870 and 1871, and Prince Edward Island in 1873, publicists and politicians found that a vocabulary and a medium – easily conveyed concepts and widely distributed journals – were available to announce the event and to celebrate the qualities of a new country. Communication networks, like trade and administrative and military considerations, were essential to the translation of place from the language of land spoken by people like the Goudie family to the language of nation that was second nature to those such as Phyllis Knight.

The great majority of the common people would have to learn to read and write if they were to operate effectively within the new dimensions. In the opening decades of the nineteenth century, parents in Upper Canada (Ontario) had used schooling as and when they saw fit, perhaps putting a child into the hands of a local tutor or academy for a few months from time to time when other commitments relented. Common schools multiplied in the 1820s and 1830s, however, and in the census of 1861 over 90 per cent of the residents of Canada West (Ontario) province claimed the ability to read and write. Still counties on the pioneer fringe, such as Algoma in the north and Prescott in the east, where families worked with axe and plough rather than with machine and account book, had literacy rates in the range of 50 to 70 per cent – a measure of the relation between the exigencies of daily life and such intellectual tools. In other parts of Canada, especially where household production of commodities continued to be the rule well into the twentieth century, nearly universal literacy arrived only slowly. Outside Montreal and Quebec City, French Canadians in Quebec joined the move to literacy around the mid-nineteenth century, but as late as the 1891 census the literacy rate in rural areas did not exceed 60 per cent. The rate in Newfoundland fishing villages and prairie fur-trade districts was lower. Thus most Canadians became literate (thereby acquiring access to the new vehicles of communication heretofore possessed chiefly by the elites) only during the nineteenth century and the opening decades of the twentieth.[8]

With the development of a literate community, there occurred an expansion in the *scale* of ordinary society, whether measured by the emergence of new communication systems, by the increase in population, or by the growth of gigantic organizations. Expansion bred crisis. The stress was the result in part of the rate of change, but also of the extraordinary sweep and novelty of the problems themselves. As Raymond Williams has argued, 'The crisis in modern communications has been caused by the speed of invention and by the difficulty of finding the right institutions in which these technical means are to be used.' The nature of the newspaper – who decided its content, how it would be paid for – became a concrete embodiment of this puzzle.[9]

The giant, privately owned, daily newspaper soon dominated public communication. Politics placed first in its affections, economic material second, and the social context of daily life, including sports and fashion as well as sensational events, third. Every newspaper created a 'map' of the land. For the dailies, this map centred on the city, the province, and Canada as a whole; the 'mother country' (Britain or France and their empires), and 'our peaceful neighbour' (the United States), came next in importance; then came the other countries of Europe; the rest of the world appeared rarely. A newspaper's selections and biases reconfirmed stereotype, status, and priority; and its choices legitimated certain values and authorities. Decisions on content and viewpoint – the actual burden of the news – represented an 'exercise of power over the interpretation of reality.'[10]

One might be forgiven for assuming that the big-city dailies set the standards for and dominated the content of all

Canadian journalism. However, copies of two journals per week were printed on average for every Canadian family in 1890 and four journals per week per family in 1930.[11] What else were people reading? In the shadow of the mass media, another journalistic universe – the world of weeklies and monthlies – bubbled and boiled. Although they may not in any single case have reached many readers, the weeklies collectively entered as many households and exerted at least as much influence – perhaps more – on the thinking and behaviour of their readers.

Nearly half of Canada's 1800 printing and publishing firms produced local weeklies at the close of the 1920s. These journals defined by ethnicity, faith, or place imagined the world not in terms of consumption or politics but as networks of families and acquaintances. The local weekly's central myths (where myth is defined as 'a complex of symbols and images embedded in narrative') provided several points of reassurance: that a particular small community continued to flourish; that it adhered to unvarying rhythms from year to year; that inherited social conventions – visiting with friends, the summer sports day, the bridal shower and the church wedding – still were respected; and that identifiable families still earned and lost status according to the rise and fall of their educational, moral, and material fortunes. The individual reader would absorb the local weekly's myths in the same manner as he or she would those presented in the urban daily. Both types of journal imagined worlds that were believable and complete in themselves. The fact that the two worlds differed utterly, and seemed to operate within different dimensions of time and space, seemed not to disorient the reader. The contradictions represented the varieties of ways in which readers adjusted to the changing – and overlapping – dimensions within which they lived.[12]

The *Grain Growers' Guide*, to take just one illustration, acted as a people's forum during the first decade of its existence on the prairies, from 1908 to 1917. Its women's page enabled readers to compare notes on how to build liveable homes and communities. The key topics in these women's discussions of cultural expression included 'four major areas of creative interest: house design, interior decoration, flower gardening and clothes.' The authors' primary concerns lay not with consumption or status or novelty, which would have dominated the urban daily's commercial advertising, but with aesthetics and practicality: 'It is my firm conviction,' wrote the women's page editor, 'that every ugly, inharmonious house in this country could be at least reduced to harmony, if not made positively beautiful, if the owners were prepared to make a lavish expenditure of energy and a small expenditure of money.'[13] The editorial-page cartoon played an equally powerful role in establishing the message that farm families (clean lines, straight jaws, plain overalls) were locked in a deadly battle for survival against the vested interests (portly men, business suits, watch chains, and cigars). Both images depicted a farm family able to adapt to difficult material conditions and eager to use new tools to improve its circumstances.

Not only newspapers facilitated these national conversations. New approaches to popular education and communication also became effective in the early twentieth century. A remarkable generation of social reformers, including E.A. Corbett (adult education), John Grierson (film), and Graham Spry and Alan Plaunt (radio), took the lead in new cultural agencies and, along with thousands of others, employed magazines, books, political manifestos, adult education classes, radios, and educational films to build a Canadian version of print-capitalist society. Their energy, commitment, and success shaped the country's social and political arrangements for the next half-century.[14] This is the debate that Phyllis Knight in effect entered during the 1930s and 1940s. She sat on its periphery, but she, with the rest of the 'plain' people, helped to inform the choices of the country. The entire edifice presumed popular ability to participate in the national discussion. Politics relied on citizens' literacy and citizens' use of the communication media.

III

In this new order, Canadians debated the relative merits of capitalism and socialism. Which was to become the organizing principle of the economy? According to the proponents of the two contending approaches, either might have prevailed – at least, that was the view in 1905 and 1935. From today's perspective, however, the struggle looks quite different. The shift reflects the post-1988, post-Soviet world, wherein capitalist institutions seem to sit astride the globe. From today's vantage point, one might interpret the ideological debates of the first half of the century as struggles over the nature of the capitalist system itself.

The cultural context of ordinary Canadians changed with the growing scope of the market and the revolution in communication. The multiplication of print vehicles required that common people become literate. The expansion of corporate organizations pushed them to build their own vehicles of communication. And the pressures of daily existence encouraged them to develop their own intellectual maps of the new age. A different challenge now confronted them: could the farm family in Alberta or the timberworker's household in northern Ontario, the longshore workers' families in Vancouver or Montreal or Saint John implement such changes? Were they able to translate ideas into political action?

Phyllis Knight did not contribute directly to the evolution of Canadian politics. Yet she did participate in the cultivation of citizens' perspectives because she conversed with her neighbours and her husband, shaped the views of her son, and cultivated or quarrelled with the administrators who impinged on every citizen's life. During her interviews in the early 1970s, Phyllis devoted few words to national or local politics or to formal religion. According to her son, she 'never went to church, never came to believe in any supernatural entities, and never changed her opinion of religion.' Such persistence is, he wrote, 'a kind of triumph.'[15] As a child in need of a baptismal certificate to show school authorities, she had discovered from her mother that she hadn't been christened: 'We weren't in any church, she didn't go in for religious nonsense.' Phyllis was also suspicious of political leaders, but she never opted out of the debate: 'Of course, I never knew very much about politics and who you could rely on. Then as now there wasn't any party or group I could support one hundred percent.'[16] Nevertheless, her votes and her perceptions of the needs of ordinary citizens, when added to the reactions of others, reconstructed Canadian society and government during the crucial years when Canadians were struggling to adapt to the new print-capitalist order. In these ordinary ways, her perspective helped to shape the nation.

Phyllis's approach to the wider society in which she lived was complicated by her German origins and her unusual domestic circumstances as well as by her personality. Her numerous changes of address during the 1930s and the frequent absences of her husband after the Second World War must have placed her on the edge of neighbourhood and school and within the sights of those who gossiped about outsiders. She did not always swallow her anger at slights. Having endured awkward moments as a day domestic worker in the early 1930s, she responded aggressively to one difficult employer who was cheating her of her wage: 'Families with little white collar jobs puffed themselves into high and mighty burghers ... So I took this woman's prize soup tureen out of the china cabinet and dropped it on the floor.' At the Oddfellows' Hall, a few years later, she found a brooch but was treated rudely by the owner who came to retrieve it: 'She was insulting and snooty as hell. If she had asked me civilly I would have given it to her right there and then.' Phyllis dropped it down a sewer. Both moments can be seen as unnecessary or dishonest, given the conventional view of property and work discipline. Seen from her perspective, however, they look more like acts of integrity in which she answered exploitation and rudeness with resistance and sabotage.

How to analyse the structure of the society in which Phyllis found herself? Her observations on the Depression of the 1930s and the boom of the 1940s suggest her approach to public affairs:

That whole life, the whole way people thought and felt and acted during the depression ended when the Second World War started. Actually, all the resources were there. The money was there, or could be created or freed. All the factories were there. They all started up pretty soon after the war began. But for ten years everything had been paralysed and according to the big shots there was nothing which could be done ... For ten years they couldn't find a way of providing jobs for people and getting them enough to eat or a place to live ... Of course, you can easily begin to feel helpless if hard times last too long. It's almost impossible not to. You begin to think, 'What can a single person do about it? You can't do anything.' Your thoughts are about how rotten it all is. Most of our energy and thought was concentrated on scraping together enough to keep a roof over our heads and getting barely enough to eat.

Phyllis did not often come directly in contact with what she called 'officialdom.' Even when Rolf was in school, she had little experience of local politics: 'Parents didn't involve themselves with schools. There weren't any parent–teacher associations or anything like that. In grade school, the school nurse always came around once or twice a year. If the child missed a number of weeks of school she would come around to see what was wrong. Those nurses were all pretty good, sensible, and down to earth people. That was about the only contact I had with the schools.'

Phyllis's sense of Canadian politics was best summarized in her comments on her reaction to Canada itself:

It's hard now, after so many years with so many changes, to keep straight what we knew and felt then. Toronto was a relatively small and provincial town compared to what I was used to. There was at least the beginnings of social security in Germany ... Here there was absolutely nothing. You have to remember that things like [Workers'] Compensation, old age pensions, and welfare or any kind of social security only started slowly around the Second World War in Canada.

Then too, we were used to a long tradition of socialism in Germany. There were millions of people who were socialists, some for two or three generations. But here, socialism was still thought of by some people as some crack pot fantasy or some foreign plot. The churches really had a big say here in Canada

What I think I really missed the most, and was most astounded not to find, was lack of social [i.e., cultural] facilities here in Canada available to ordinary people. Of course there were plenty of people here, and in Berlin, as everywhere who never did anything except visit their families, talk about their work and children, and go to their local pubs. But in my day, in Berlin, as in the other big cities of the world, there had developed all sorts of ways for ordinary people to have as full a round of plays and lectures and what not, as they wanted. Museums ... opera and theatre ... zoos ... well stocked public and loan libraries ... There was very little of that here. And that really did shock me.

In the late 1930s, Phyllis and Ali participated in some of the Vancouver workers' political activities, including an occupation of the post office by the unemployed, some May Day events that culminated in speeches in Stanley Park, and the dispatch of troops to fight with the Republican forces in Spain. She said that she didn't know much about politics; nevertheless, her reactions to such events, and to wider economic conditions, actually helped to shape the character of Canada's reform movements.

One measure of the changes was Phyllis's own circumstances. By the end of her life, with the government pension and the medical care system, she actually felt secure: 'It's the first time in my life that I have enough and don't have to count every penny or worry how long it's going to last ... Oh sure, I'm grateful enough for the pension and medicare, but then we earned it. So what's to be grateful about, really? The profiteers rake in the money by the hundreds of millions like always.' The welfare state had made a significant difference to her last years. Her own struggles, along with those of thousands of other ordinary people, had reshaped the economic circumstances of everyday citizens.

IV

The history of citizens' involvement in Canadian politics is a story of victory as well as of defeat. In the battles that they waged, and in the moral victories that they claimed despite setback after setback, the common people did change Canada's version of capitalism. According to the national political interpretations of history, their greatest victories lay in the political institutions that evolved. But their gains also included, to mention just a few examples, reform of the prairie agricultural economy, of the national system of industrial relations, and even of the range of choice among political parties.

The story begins with democracy itself. The struggle for universal suffrage in Canada lasted nearly one hundred years and for most of that century did not look like a story of progress. The dominion franchise act of 1885, which ensured that all males could vote, was accompanied by the disfranchisement of most Aboriginal Canadians. The first group of Canadian women obtained the right to vote in federal elections in 1917, but many immigrants from central and eastern Europe lost the franchise for that same election. Quebec women could vote in provincial elections only from 1940. Chinese and Japanese Canadians won the franchise only after the Second World War. And Aboriginal people voted in Canadian elections as a matter of general right only beginning in the 1960s. Nevertheless, common peoples' political narrative should be seen as a success story because it culminates in universal suffrage. The continuing challenge – and a much more difficult battle – has been to ensure that citizens' votes are informed and effective. In this, today's Canadians could learn a lot from and owe a debt to previous generations.

Though an exclusively male sphere, Canadian politics belonged to both the people and the elites in the late nineteenth century. The party as institution and the two-party framework of policy choice became the norm between 1850 and 1900. To participants and onlookers, electoral politics must have seemed a cross between men's athletic contests and men's theatre. Sir John Willison recalls in his *Reminiscences* a political meeting in 1872, his first experience of a Canadian election, in which three candidates spoke for hours, rallying the crowd with emotional pitches, hurling scorn at their opponents' arguments, and offering the equivalent of the preceding generation's revival meeting.[17] Sir John A. Macdonald, Conservative party leader until his death in 1891 and prime minister from 1867 to 1873 and 1878 to 1891, conducted an extraordinary correspondence with Canadians in all walks of life. His was still a kind of people's political party, whatever the relation between its policies and the interests of particular groups of entrepreneurs.

Formal political institutions grew more remote from the people in the opening decades of the twentieth century. After four men had failed to reconstruct Macdonald's empire, the next Conservative leader comparable to Sir John A. in influence, Robert Borden, put together a party machine run by specialists in electoral organization and policy-making. His correspondence, much of it with provincial and regional party leaders, reads like the work of an engineer, full of business and practicality.[18] Business leaders provided the largest proportion of campaign funds spent by the two major political parties, the Liberals and Conservatives. Party professionals dedicated themselves to cautious management of the nation and, more precisely, of its economic institutions.[19]

Common people responded effectively to the elite's political and economic arrangements. Their replies, at first piecemeal and fractious, later carefully considered and country-wide in effect, addressed the excesses of this system. Their goals included a living income, a measure of control over the pace of change, and some influence over the direction of the entire society. Phyllis belonged to their movement.

To follow the change in public policy, one must understand the atmosphere in the dominant cultural institutions. Phyllis Knight was correct in her assessment of the place of religious institutions in Canadian public life: the churches did exert great influence during the first half of the twentieth century. The change in their outlook from the preceding century was, however, especially significant for the politics of English-speaking Canada.[20] In the first half of the nineteenth century, many of the ordinary people of English-speaking British North America discussed the world around them in the language of evangelicalism and, when economy or business was involved, in the language of self-interest and family gain as well as of reciprocity and charity.[21] Their evangelicalism was a religion of popular liberation. More a loose creed than a consistent theological system, it emphasized psychological intensity, individual conversion, the reality of sin, and the necessity of redemption through God's grace.[22] This perspective survived for three or four generations and represented a kind of consensus for many anglophone Canadians in the years when the nation itself was taking form. The consensus embraced the four largest Protestant churches (Anglican, Baptist, Methodist, and Presbyterian) that together consti-tuted 'an unofficial moral and educational establishment' among English-speaking Canadians. This ideological agree-ment legitimated the new capitalist order *and* the personal values – thrift, sobriety, work discipline – that would ensure its hegemony.[23]

This theology lost relevance after the turn of the century because the old language was no longer sufficiently flexible and evocative to articulate the philosophy that dominated in the new order. Evangelicalism was too individualist and was better suited to a society of small towns, family enterprises, and face-to-face relationships. The very scale of communications in the new age required new forms of control. The social realities of large cities and governments, like large corporations, demanded greater attention to collective strategies. Thus evangelicals entered a period of questioning in the decades before the First World War and departed from the 1920s in several different, mutually opposed groups.

By the early twentieth century, one major new strain of religious language – the social gospel – challenged dominant assumptions about relations between churches and public policy. The social gospellers called for the salvation of society itself. For them, the Promised Land was no longer in an otherworldly Heaven but had descended to earth. They saw the land in which ordinary people dwelt as synonymous with both Heaven and Hell, and these common people – many of whom lived closer to underworld than to utopia – wanted immediate reform of immediate evils. As evidence of the extraordinary and continuing power of the churches in Canadian society, the social gospel imprimatur was said by many observers to forge 'links between proposed reforms and the religious heritage of the nation, in the process endowing reform with an authority it could not otherwise command.'[24] Reforms undertaken in the social gospel's name included immigrant settlement houses in urban slums, campaigns for cooperative marketing institutions in the Maritime fishery and prairie farm districts, and a new place of worship for working people called the Labour Church. It also sustained a new national Protestant movement – the United Church of Canada, formed in 1925 from Congregational, Methodist, and many Presbyterian churches. These activities, some of them undeniably secular, displayed changing perceptions of individual redemption and community salvation.[25] During the interwar years, as such debates continued and the gulf between the camps grew, the social gospel gave way to factions of traditionalists and progressives in the churches. Perhaps more important in the longer term, it also led to a powerful, secular, political-economy approach to social and cultural issues that can be called Canada's version of the international doctrine of socialism.

Canada's distinctive socialism deserves special attention because it provided intellectual coherence to common people's view of the world during the 1930s and 1940s and in fact for the following generation. In prairie farm associations, in the Antigonish movement of the Maritime provinces, in country-wide trade union and adult education activities, in the Canadian Broadcasting Corporation and the National Film Board and artists' campaigns for a national culture, this evolving, homegrown version of socialism became an alternative to the pure capitalism of markets and competition that was running aground on the Great Depression and the Second World War. Its character was explained by J.S. Woodsworth, first leader of the Co-operative Commonwealth Federation (CCF), to the party's founding convention in 1933: 'Undoubtedly we should profit by the experience of other nations and other times, but personally I believe that we in Canada must work out our salvation in our own way. Socialism has so many variations that we hesitate to use the class name. Utopian Socialism and Christian Socialism, Marxian Socialism and Fabianism, the Latin type, the German type, the Russian type – why not a Canadian type? ... We in Canada will solve our problems along our own lines.'[26] These were the sentiments that Phyllis Knight encountered in Vancouver during the 1930s and 1940s.

The CCF, the Communist Party of Canada (CPC), the cooperative movement, the trade unions, adult education groups, and the League for Social Reconstruction would have vigorously denied it, but, despite obvious differences, they spoke a common language. They defended the welfare of the community against special privilege, rejected the profit motive as a foundation for human relationships, proposed that planning should supersede competition in order to eliminate waste, and defended the ballot and popular decision making. Neither British nor American in approach, though informed by the thinking of radicals in both countries; not simply Marxist, though informed by Marxist perceptions of a capitalist age; not always religious, though redolent with biblical imagery and evangelical language and inspired by concern for the welfare of others: this was Canadian socialism, successor to evangelicalism and social gospel as the alternative to capitalism's excesses.[27]

If Canadians have been slow to recognize the socialist forces that shaped their recent history, it may be because the perspective was expressed through a number of quite different institutional vehicles. Reformers in church, farm, intellectual, and labour groups never agreed to throw their support behind just one political party. Some of their energy sustained the CPC and the CCF (from 1961, the New Democratic Party, or NDP). Some was infused into the Liberals. Even Conservative and Social Credit governments accepted aspects of the socialist analysis, especially when it was cloaked in the populist rhetoric of a William Aberhart in Alberta, a Maurice Duplessis in Quebec, or an H.H. Stevens in Ottawa. Thus socialism pervaded public policy debates for the next generation. It shaped the Liberals' acceptance of Keynesian economic policies, of collective bargaining in industrial relations, and of such social welfare measures as the family allowance and unemployment insurance. It also informed the CCF's pioneering of hospital insurance and state-administered medical care in Saskatchewan. But because the Liberals controlled the federal government for fifty-one of sixty-three years between 1921 and 1984, they legislated most of Canada's version of the welfare state.

Socialism was less influential in Canada than in continental Europe, Britain, Australia, and New Zealand. However, the Canadian movement was stronger than its U.S. equivalents. There, Communists, Progressives, Socialists, cooperators, and unions had also tried to push legislators to adopt socialist solutions. The consequent legislation, which owed its existence to the administration of Franklin Roosevelt, had been achieved over bitter opposition, however, and only because of the devastating Depression experience. The business back-lash thereafter was fierce. In Canada, although cooperative and farm and labour groups had rarely agreed even on the time of day, a single political party, the CCF, did consolidate these Canadian socialist voices at the national level and became an alternative government in some provinces.

The continuing existence of Canadian socialist reformism stood out in North America. By the late 1960s, Canadian blue-collar and farm families had built collective institutions of self-defence similar to those earlier legislated in Britain, Germany, Scandinavia, and Australia, among other countries, in earlier decades. While acknowledging the relative slowness of the process, in comparison to European or Australian precedent, observers of public policy must also recall the circumstances of North American society and the relative success of these Canadian movements within a North American context. Henceforth Canada's common people possessed institutions – be it credit union or *caisse*, co-operative store, health insurance, pension plan, political party, union – that reshaped capitalism. Their spokespeople offered a made-in-Canada version of the international welfare state that differed substantially from the next-likeliest alternative – an American-style, more market-oriented system.

Between the mid-nineteenth and the mid-twentieth centuries, the spread of print and of Canadians' ability to read and write sustained a frame of mind that could be described as increasingly 'national.' At the same time, Canadians' immersion in the practices of modern industry and trade represented a local version of this global transition. Newspapers and railways – products of the machine age and creators of so much of the country's imagery – exemplified the new pace of cultural exchange.

The economic forces of print-capitalism increasingly drew ordinary Canadians such as Phyllis Knight into a *single* community. In this third Canadian construction of the dimensions of space and time, factories and farms and offices provided the economic settings in which ordinary families made their living. Rather than region or religion or ethnicity, wages, property values, and prices increasingly measured common people's daily existence. Citizens passed their time increasingly within the common units of monetized minutes and work weeks and productive careers. Within a century, the space in which they lived – once a backwater of small, scattered, poor colonies – had become a transcontinental nation. The enormous change in the dimensions of time and space, as experienced in the daily lives of common people, demanded extensive adaptations. These ordinary citizens did adapt, forging new institutions with their literacy-based skills, their weekly newspapers and adult education programs, and their socialist analysis. Common people's cultural responses, as Phyllis Knight's life shows us, rebuilt Canada.

PART FOUR

Screen-Capitalist Societies

7

Roseanne and Frank Go to Work

Canadians today perceive the dimensions of their environment differently than did Alestine Andre's grandmother, the young Elizabeth Goudie, and the young Phyllis Knight. They see the world so differently that some historians describe their situation as a fourth version of time and space. This new and distinct context of communication exists for reasons that are as obvious as the computer, satellite, television, and cable technologies that surround us. But the new technologies have also created a world in which the leaders of the great social institutions seem to hold *all* the power.

How are common people helping to develop Canadian institutions today? Are they losing the race to control their own destinies? It is much too soon for ordinary citizens to have recorded their reflections in a manner similar to the manuscripts written by Elizabeth Goudie and Phyllis Knight. After all, the lives of these children and young adults commenced in the 1960s and 1970s. Thus chapters 7 and 8 rely on interviews that I conducted with one individual, on Ken Dryden's unusual book about an Ontario man, and on a remarkable exception, a four-volume memoir prepared by a Quebec woman.

What one should call this contemporary era is a subject of debate. Some observers speak of globalization as if this was the first society to be affected by globe-encircling phenomena; in contrast, I have suggested that this distinction belongs equally to the preceding epoch. Others speak of 'information society,' 'knowledge society,' and 'culture of consumption.' Yet others, and there are many scholars in this camp, suggest that a former age has ended but are debating vigorously how to characterize the new one: they say that we are living in a 'post-' moment, meaning that they are thinking in postmodern, postindustrial, post-print, or poststructural terms. Rather than engage directly in these debates, I am going to make do with 'screen-capitalism,' a phrase that unites economic and communication forces in one term, just as oral-traditional, textual-settler, and print-capitalism, which I used above, have done. I believe that Canadians have embarked on a fourth phase in the history of their community's dominant communication institutions and that the response of ordinary people to the new circumstance is slowly taking shape.

Two features of the contemporary epoch are particularly noteworthy: first, the exceptional alterations in the perceived dimensions of time and space, a result of shifts in communication and transportation systems; and, second, the unprecedented juxtaposition of insecurity and plenty. These two economic themes constitute the subject of this chapter. In the next, I turn to the cultural and political aspects of our contemporary age.

I

Roseanne is over forty and entering middle age. She lives in a large city, not a metropolis but, like several dozen other Canadian cities, one of the next tier. Her house sits among hundreds of others in a typical suburban tract that is snowy in winter, green in summer. She and her husband are raising four children (one from their recent marriage and three from her earlier relationship) and, although they are not wealthy, she describes their economic circumstances as relatively comfortable and certainly far more satisfactory than those of her own childhood.[1]

Frank, in his mid-forties when author Ken Dryden told his story, is a married father of three. From the outside, his life is similar to that of Roseanne, although he lives in a much larger community, a suburb of Canada's largest city, Toronto, and works for a much larger organization, the petroleum giant, Imperial Oil. Frank values the swimming pool in his backyard, his total repayment of his mortgage, and, most of all, the solidity of his family life. Frank and Roseanne can be taken as illustrative of the changes that have moved ordinary people in new directions during the decades since the Second World War.[2]

Roseanne grew up in a small, poor settlement near an Aboriginal reserve. Its principal mode of transportation remained horse and wagon until the early 1950s, but her family had a radio and she attended the Friday night movies in the village hall. She spoke her mother's Aboriginal language almost exclusively in her early years, but the public culture and rules at school ensured that English became her main language of expression after the age of five, and she says that she now speaks only English.

Roseanne's tiny household may have seemed prosperous compared with that of many of her neighbours, but ten children stretched the resources of its two-room dwelling to the utmost. She was delighted when her parents decided to move to the city and to exchange their crowded shack for a larger house. However, the transition, which occurred when she was ten, became a disaster for her. She encountered racism in the schoolyard. Her father's business failed. Her mother ran into battles over welfare support. She and her brothers and sisters knew what it was to have headaches and empty stomachs at the end of the month when they literally went to bed hungry. She managed to graduate from an alternative secondary school, one bright moment in those difficult years, but the dark times, as she sees them, between her early teens and early thirties included constant changes of residence, the birth of three of her children in an unstable, unsupportive relationship, and, finally, the rejection of her first desperate application to a university program designed for promising young people in straitened circumstances. However, admission to the program on her second try provided the chance that she needed. She became a teacher, moved from challenge to challenge in her occupation, and now thinks of herself as having achieved a reasonable degree of integration within what she calls 'mainstream society.'

Frank came from a more prosperous home, not wealthy but comfortable. Although he benefited from his stable family setting, his dyslexia was not recognized at the time, and he had trouble in school. Never the star student or athlete, he travelled on the fringe of the action, struggling with assignments, preferring his waged work at a grocery store; but he stuck with school until, at the age of twenty-two, he completed grade thirteen. He married two years later, found a steady job, and stayed with the company for twenty years, shifting slightly from one position to another while his children grew up. Ken Dryden chose him almost at random to illustrate the lives of those Canadians 'who go unnoticed because they aren't very noticeable and who for the most part don't want to be noticed.'[3]

Roseanne and Frank resemble Elizabeth Goudie and Phyllis Knight in the obvious ways. They grew up in family settings where chores had to be done. They made their own way as young adults, finding jobs and building families despite considerable challenges. Now, as they enter middle age, they seem to have made their peace with the wider world. Roseanne and Frank differ from Alestine's grandmother, the Labrador trappers, and the Vancouver wageworkers in the relatively greater material comfort of their modern suburban tract homes, their continuous work-related, print-based education, and their apparent integration within a national and international community. Yet they are the inheritors of the centuries of experience in this land negotiated by Aboriginal, settler, and capitalist households.

Despite her apparent prosperity, especially in relation to the Knight and Goudie households, Roseanne does not feel secure. She says that she belongs in the insecure fraction of the Canadian population. The designation may seem inappropriate, but it carries authority. On the basis of a widely cited and very large poll conducted in 1994, the Ottawa-based Ekos research group divided adult Canadians into five categories: 'elite insiders' (19 per cent) and the 'secure middle' (24 per cent) accounted for less than half of the national population; the 'insecure middle' (16 per cent), 'angry and alienated outsiders' (19 per cent), and 'economically disengaged dependents' (22 per cent) made up the remainder. Those who believed they controlled their lives, at 43 per cent, were fewer than those who did not, at 57 per cent.[4] Roseanne and Frank belonged in the larger, insecure fraction of Canada's population.

According to the sociologist Wallace Clement, advanced capitalist societies now exhibit distinctive occupational patterns associated with the individual's apparent control over the means of production and over work obligations. Clement distinguished capitalists and senior managers from working people, as earlier students of society have done, and then defined two additional groups, both in the middle class but differing in their ability to control their own labour and the labour of others. According to Clement's calculations, nearly 60 per cent of the workforce in Canada belonged to the working class. Just over 10 per cent belonged in the old middle class, and another 25 per cent in the new middle class. Many of these people did not control their own labour or that of others. About 6 per cent were in the capitalist/executive class. Clement's definitions suggest the continuing preponderance of workers – perhaps two-thirds or more – who do not own means of production and do not possess the power to control their own and others' labour. Members of the latter category, like Roseanne and Frank, were the ordinary citizens who had to maintain a sense of balance despite constant challenges.[5]

Common people recognized their location within society. Roseanne divided the community into four parts: the wealthy quarter, the poor quarter, and two quarters in the middle. She believed that her own household had recently moved out of the poor and 'just over the border into the middle,' but her mother remained in the poor fraction. Like Frank, Roseanne was cautious in her view of household security. Although she lived in a two-income household, she wondered whether she and her partner – and, more compellingly, her children – would be able to maintain the relative comfort that they now enjoyed. Amazingly, given the level of her household income, she was less secure about her economic standing than Phyllis Knight and Elizabeth Goudie had been. Why the paradox?

II

Work remained central to the lives of most adult Canadians under the age of sixty or sixty-five at the start of the twenty-first century. A close analysis of their daily labour revealed aspects of the task that each person could take pleasure or pride in and aspects that were far from fulfilling. The work itself, however, was very different from the labour under-taken by the Andre, Goudie, and Knight families.

Both Roseanne and Frank were conscious of time and extremely disciplined in their response to its signals. Both made a point of arriving not just on time for work but early. Both expected others to be similarly punctual. Not surpris-ingly, time became a matter of some tension in the disciplin-ing of their children. What had been unknown in traditional Aboriginal society, irrelevant to Elizabeth Goudie, and resented by Phyllis Knight – the daily passage of measured, monetized minutes – had become so internalized in their lives that they could not imagine not being subject to the dic-tates of the clock.

When asked about the cause of daily anxieties, workers consistently mentioned the pressures of time. These increas-ing demands were evident at Imperial Oil. Ken Dryden watched Frank at work:

It was worse a year ago when he was new to the job and knew
he couldn't do it ... Everything moved too fast. He couldn't
read his screens, sort them out, understand, formulate, make
good decisions and know they were good. The pages kept
turning before he was done. Every decision seemed reflex,
knee-jerk, random ... Now, he speeds through his work-lists,
cuts corners, knows more, gets better, yet every morning his
race ends before the finish line is in sight. This afternoon he'll
have to go back to these lists, he promises himself, but that
won't happen. Not many months ago he just waited for the
sound [of the phones that announced incoming work], and
when it came, he could feel his body scream. Now, there's just
a tiny scream.

In a Hamilton steel plant in the early 1980s, a new management team wanted to streamline the production process,
reduce the number of workers, and simplify the actual tasks
carried out on the shop floor. The pressure to get more out of
each worker fuelled tension during the workday and
inspired some graveyard humour. The workers labelled the
plant's chief executive 'Chainsaw Bob' or 'Neutron Bob' ('he
destroys people but leaves buildings intact') and said that
the supervisors, who had to ensure that the new directives
were carried out, had been trained in 'Charm School,'
although the managers themselves would have explained
that they had studied 'interactive management.' Their goal,
as Edward Deming's approach to 'Quality' would dictate,
was *constant* improvement in work practices, a seemingly
innocent and appropriate target unless one contemplates
what it might actually mean if pursued forever in the conduct of relatively uncomplicated tasks.[6]
Few workers ever enjoy the prospect of radical restructur-
ing of the workplace. Yet constant change for the purpose of
greater profitability and productivity is becoming a normal
aspect of life in most jobs. Neither Alestine Andre's grandmother nor Elizabeth Goudie ever encountered insistent,
externally monitored, timed changes in their daily work. In

contrast, the threat of change on the job made Frank and his co-workers very uncomfortable. Imperial Oil merged with Texaco while Dryden was writing Frank's story, and the office staff of the two corporations had to endure months of uncertainty about 'downsizing.' 'It is 7:55. His work day doesn't begin until 8:30, but already he's at his desk. He can't be sure of traffic on the expressways so he takes Sheppard Avenue and Don Mills Road, and not sure of them, he leaves early and ninety-nine days and more out of a hundred, arrives early. And after a few smiling hellos and comments about the weather, with nothing else to do, he begins early. Around him, co-workers still sip their coffee and chat, about the Leafs who have just begun another NHL season, about the baseball playoffs, but mostly about the "rumours." Two thousand of them are going to be laid off, four thousand some have heard.'

These workers had become accustomed to precise clock-timed labour and to continuous significant changes in the work process. They dealt daily with messages, print and oral, that conveyed new job requirements and attempted to inspire a more urgent commitment to a common enterprise. And, like Frank, they faced an inescapable uncertainty that could cause them sometimes to awaken in panic in the middle of the night. This was not the insecurity of the Aboriginal hunting family in the face of a food crisis. Nor was it the insecurity of Elizabeth Goudie's family in a period of shortages, since she and Jim remained confident of their skills and knowledge. Nor yet was it the Knight family's round of wage-labour assignments, since, in their case, there was always another job to replace the one just abandoned. For Frank and Roseanne, this was insecurity for as long as the work career lasted, and it was driven by challenges to one's work performance and by instability in the workplace.

Roseanne and Frank are representative of Canada's common people today not just because they are responding to typical time and workplace pressures but also because their occupations are part of a new type of economy. The number of Canadians engaged in farming declined sharply after 1945. And those in fishing, trapping, forestry, quarrying, mining, oilwell drilling, and other primary industries had become so few that, taken together with the farmers, this category constituted less than 10 per cent of the entire workforce by the 1990s, compared to nearly 30 per cent in 1941 and 45 per cent in 1901. The proportion of Canadians living on farms had fallen below 5 per cent, the proportion living in small towns had dropped below 20 per cent, and nearly 80 per cent now lived in cities.

The habits of print-capitalism – market, property, profit, commodification – still govern the daily life of almost all Canadians. For most people, the wage continues to be the normal intermediate step between their labour and their daily bread. However, as Canadians moved to urban centres and as the characteristics of urban culture invaded the countryside, life associated directly with the land in the way that families reliant on farm and fishery and forest had known it in earlier generations, including the Goudies and the Andres, nearly ceased to exist. At best, life on the land for most Canadians became a part of urban rhythms, fitted into metropolitan calendars in the form of holidays, school outings, and work-related schools and seminars.[7] Even farming and fishing became part-time occupations, supplementary to other sources of income.

While the number of farmers and primary-sector workers declined, so did the proportion of workers engaged in blue-collar manufacturing work. Steel and auto parts and airplanes and mining equipment still required labourers, but less of this work took place in Canada – relative to the rest of the world – and, as at Stelco's vast steel plant in Hamilton, a factory's automated manufacturing processes required fewer workers per unit of output.[8] Thus Ali Knight's types of jobs in forests, rail crews, and mining camps declined as a proportion of the labour market, to be replaced by a new kind of task.

The crucial difference in work rested on the proportion of

labourers who possessed service-sector jobs – labour that involved schools and hospitals and media outlets, sales and finance and advertising, clean hands and electronic pulses and shuffled paper. This was where the work of Frank and Roseanne belonged. Such tasks increased sharply from about 40 per cent of all jobs in 1947, to nearly 60 per cent in 1967, and to almost 70 per cent in the 1990s. Frank worked at a desk, using computer and telephone constantly as he assessed credit risks and measured clients' payment patterns. He was required to reconcile the customer's personality, as conveyed over the telephone, with the financial data on his computer screen, simultaneously urging the necessity of debt repayments and of personality changes or, at the least, the promise of better behaviour from difficult credit-card holders. Roseanne taught and administered other teachers in a public school, trying to communicate the necessity of adapting to new curricula and the urgency of achieving higher test scores. Their work ranged from the complex to the simple. One of Frank's co-workers observed that she enjoyed the 'keying' or word processing in her daily work and delighted in the feeling of triumph when she deciphered some impossible handwriting and reduced it to clear print. However, after thirty-three years in the office, she also found the daily grind unattractive: 'It's so boring ... It's a nothing job. Everybody's job is a nothing job, on this floor ... Anybody can do what we do. Not necessarily well. I mean, not everybody has the patience to sit in front of this screen or talk to angry people all day.'

Frank and Roseanne did not represent adequately the hundreds of new job labels associated with the new economy. These titles began in the familiar terrain of the 'cultural industries' such as film, newspaper, magazine, book, television, radio, and recording studio. But they also included the tasks in the 'communication industries,' including computer network, telephone, cable, satellite, and the various types of data carried on these media. Add the disciplines of marketing and advertising, lobbying and polling, and art and design. Add to this the computer-based tasks associated with games, literature, conversation groups, sales pitches, and educational programs, whether in CD-ROM or on the world wide web. The labour force had changed.[9]

The dominant resource of this age, far from being physical, has been described variously as information, culture, and knowledge. Its means of conveyance might include speech and print but pre-eminently was electronic. Its means of reception obviously required the normal human faculties but also a range of machines that would bewilder a time traveller from the 1920s, not to mention the 1620s. Many Canadian jobs now involved a considerable degree of information-processing, application of specialized knowledge, and production of intellectual property. The site in which this economic activity took place also differed from that of its pre-1940s equivalents. The land as landscape, seedbed, and resource frontier – grandmother Andre's Mackenzie River and Elizabeth's Labrador – had withdrawn; in its place Canadians grappled with abstract depictions of the real world conveyed to them by electronic systems.

Space had been restructured because the communication media had eliminated so many of the inherited constraints of physical existence. What was once a single factory might be divided among ten cities or ten countries, although its final product, whether an easy chair from IKEA or an airplane from Boeing, occupied only one physical space at one time. An office might be situated in the homes of a thousand data 'inputters' whose only common contact was a distant switchboard that monitored their progress and reprimanded them if their output faltered. A face-to-face society was giving way, it seemed, to a face-to-screen society, one that enabled an organization to stretch around the world while retaining the instantaneous connections formerly available only in a single room or a single village square.

Such reorganizations affected not just individuals but also entire communities. The international dimensions of space and time had shifted again. The geographer David Harvey, in discussing the changing relations between space and time, suggests that 'space relations have been radically restructured since around 1970 and this has altered the relative locations of places within the global patterning of capital accumulation. Urban places that once had a secure status find themselves vulnerable ... [Residents] ask what kind of place can be remade that will survive within the new matrix of space relations and capital accumulation. We worry about the meaning of place in general when the security of actual places becomes generally threatened.' Citing Sheffield and Detroit as examples of cities that have lost influence (he might have used Winnipeg, Manitoba, and Sydney, Nova Scotia), he explains that developments in transportation and communication technology have made all phases of production more geographically mobile, so that 'multinational capital has become much more sensitive to the qualities of places in its search for more profitable accumulation.'[10]

These forces were far from the daily thoughts of Roseanne and Frank. While they conversed with their friends about traffic and crime and child care, technological forces that they might never hear about and corporate decisions taken in a thousand different locations were shaping the economic fate of their regions. That the very future of their neighbourhoods, the value of their homes, and the nature of their workplaces – the spaces in which they lived their lives – were affected by such decisions was not part of their calculations, but it was part of their daily experience.

III

The new age brought with it a new relation between public space – the sphere to which Frank and Roseanne assigned their work and citizenship responsibilities – and private space. It also brought unprecedented invasions of personal privacy. As a result, and unlike grandmother Andre in the Mackenzie Valley, the Goudie family of Labrador, and even the Knight household in Vancouver, citizens today such as Frank and Roseanne experienced the boundary between what was public and what could remain private as a zone of serious contests. In the public sphere, Frank and Roseanne earned income in exchange for monetized minutes. They did their best to serve the interests of the employer when they were at work, but they did not take their jobs home. The private sphere they guarded jealously. But how did they maintain the separation, and what was the implication of the polling company's phone call, the supermarket's analysis of their purchasing patterns, the political canvasser at the door, the advertisement that urged them to do or be something new?

The divide between public and private space never had seemed so contentious. Ken Dryden was visiting Frank one Friday night when the phone rang and a customer began to explain something about his gasoline credit account. Frank reacted to this invasion of his domestic sphere with bewilderment, then shock, and then anger. He seemed to be genuinely shaken by the incursion and demanded to know how the customer had traced him to his private residence. Although the explanation was innocent, the sense of invasion remained: 'How dare you call me at home! You have no right to call me at home!' Frank stewed about it all weekend. Dryden recognized correctly that he had stumbled onto something of importance in Frank's view of the world.

The line between public and private space became more contentious in the last half of the twentieth century, in part because women's waged work became so common. The sudden introduction of large numbers of women into the waged workplace must rank as one of the outstanding changes in Canadian society between the 1950s and the 1990s. Almost half of Canada's waged workforce was composed of women in the 1990s, as compared to less than one-fourth in 1956.[11] All four adults in the homes of Roseanne and of Frank worked for wages, whereas Phyllis Knight was an exception in her generation.

The pressures created by the normal activities of daily life were undeniable. Frank's wife, Carolyn, noticed the differences after she took a job: 'The rest of the family hasn't found it easy with her gone so much ... More tired more often than before, she knows that sometimes she takes out her problems on him and the kids. She can even feel it happening as it's happening, yet can't quite stop herself. They knew they would have much less time to do everything. He and the kids knew they would need to do more, and she do less; they all understood, and accepted ... So when she comes home and the house doesn't look as it should, she redoes it or yells at them ... Mad at them and mad at herself, she turns on them more and can't stop because, in part, she doesn't want to.' This ill-humour was a product of time pressures. It was also a function of insecurities bred by each family member's expectations of what was 'normal' – a concept that contained any number of assumptions and could so easily shift in response to pressures exerted in the public sphere, such as advertising campaigns built around supposedly idyllic leisure moments.[12] And it certainly blurred the distinctions between public and private worlds.

How did one assemble meaningful packages of time in a life fractionalized by part-time work, domestic responsibilities, and a thousand possible activities for entertainment and personal enrichment? Roseanne is a whirlwind: tape the television program, collect a phone message from her mother, microwave the child's dinner, find a baby-sitter, buy tickets for the school concert, and arrange a ride to the next bingo game or fitness class: it's all done in a few minutes. But note that all six acts relate exclusively to her household or private concerns. And when she heads off to work in the morning, she leaves all these concerns behind. Two worlds; two sets of pressures.

This two-chamber life, divided by a line that is constantly blurred by contests over priorities and needs, represented a new perception of time and space. There had always been a significant range of activity associated with private time and private space – dimensions arising out of one's body, feelings, and family or household. Alestine's grandmother and Elizabeth Goudie would have recognized the distinction. However, for them, the boundary would have rested between the thoughts in one's head and the communications one had with one's fellows. Phyllis Knight's view was recognizably modern, by contrast, because it distinguished among three realms – the world of waged work, the sphere of community conversation, and the domain of the household. Mrs Knight insisted on the separation of private and public but had no difficulty discerning who were the agents of state and capital – bosses, police, teachers, public health nurses – and confronting these potential threats to her domestic peace. For Frank and Roseanne, the wheel had turned again. Surveillance had become a fact of daily life in both public and private spheres, and, what was more, it was harder to know when one was alone and when one was being watched. Supervisors at work, video cameras in the mall or the lobby, computer records of purchases and movements all bespoke an increasingly wide range of monitoring instruments.[13]

The reasons for the increased surveillance rested partly with the nature of the new jobs and partly with the capacity of modern technology to watch the worker. Some jobs offered greater freedom, and others were monitored closely. Fortunate citizens acquired skills that permitted their entry into a privileged labour market promising long-term employment, good wages and benefits, perhaps even fulfilling work, and small amounts of external surveillance. The less fortunate earned livelihoods in places where wages were lower, benefits were fewer, job-changing was common, the work itself was less likely to be pleasant, and surveillance was insistent. If they were making hamburgers, caring for children, or clerking in a store, they might work only part-time and might have little or no union protection. They could even be required to smile pleasantly at customers – a regulation that could be not only imagined but also monitored. Roseanne's children, now in their teenage years and ready for such jobs, were learning the ways of this part-time labour market.[14]

The changes in work, gender roles, and surveillance technology altered the dimensions of space and time. They also introduced new complications to family life and to one's self-perception.

IV

Were ordinary citizens more secure than their predecessors, or less so? To visit Roseanne and Frank at home was to encounter a vision of abundance. Toys overflowing the games room, clothes overflowing the closets, abundant food in cupboards and refrigerator, space for every family member, shelter of a remarkable quality, and, wonder of wonders, a beautiful pool in Frank's yard that would have seemed a plaything of the very rich when Phyllis Knight was young. Was this not clear evidence of prosperity and security? And yet one must somehow catalogue as well the thousands of messages that entered their houses, and minds, every day – too fat, too plain, out-of-date, out-of-fashion, get a car, take a trip, get smart, get happy. Who could stand unbowed under such a shower of complaint and enticement?

Let us start with the abundance. Its presence in the lives of Frank and Roseanne was undeniable. They owned a lot of material goods and possessed what used to be called creature comforts. Canadian household income, to which women's work contributed an increasing proportion, was relatively far higher by the early 1970s than it had ever been before. Although such indices are notoriously unreliable, economists declare that, after one extracts the effects of inflation, the average worker's income actually doubled between the mid-1940s and the 1970s. It was no wonder that many Canadians seemed to have more disposable income, that they purchased new houses and paid for vacation travel and additional cars. Their household incomes did not rise appreciably in the next two decades, however, and relative stagnation bred further insecurity. But the willingness of more family members to enter the workforce usually enabled the household to continue its previously established consumption habits.

To reinforce this perception, recall the victories of Phyllis Knight's generation in controlling the excesses of capitalism. The arrival of the welfare state in Canada, like the increase in women's work for wages, helped transform the organization of 'plain' people's lives in the second half of the twentieth century. If working people experienced losses as well as gains in the workplace, they none the less found that in post-1970 Canada they could make ends meet. One crucial reason was the increasing amount of part-time waged work available for some members of the household, including children and women; another was the new support available from institutions of government. In this era, the historic link between work and income had actually ruptured. For the first time in history, people did not have to expect absolute disaster – as in the cruel Victorian workhouse or starvation on a medieval peasant acreage – if their health or ability to earn a living declined. Wage-earning families had been hostages to fortune in the first century of Canada's industrial age, just as their ancestors in peasant societies had been, if they became ill or sustained an injury.

Between the 1940s and the 1970s, however, conditions did improve. Even if working families did not secure lifelong employment, they could use retraining schemes, job-finding services, and, when necessary, welfare payments. They could expect insurance payments during short-term unemployment and even crop failure. If problems arose in the workplace, three of every ten Canadian wage-earners could turn for assistance to a union that had been sustained by government-created rules on collective bargaining; some of the remaining six in ten benefited, along with their organized colleagues, from the institutional arrangements for bargaining, grievances, lay-offs, safety, and training that unions had achieved. They looked forward to a government-run pension plan when they retired. Government and union had replaced church and neighbours as significant supports in the story of the wage-earning household's material conditions after 1940.[15]

Roseanne had benefited directly from one of those government schemes that offered special support to welfare recipients wanting to further their education and thereby increase their chances of gaining employment. She regarded free medical care as part of her children's birthright. She did wonder whether anyone – government or union – could make a real difference when individuals in the workplace became downright hostile, or when economic changes caused lay-offs, but she did not blame the supporting institutions as much as she regarded the crises as inevitable. Like Phyllis Knight, who had often lived on the margin of poverty, Roseanne was defiant about her sometime reliance on state support: 'We have this frame of mind ... us poor people. (I'm not even poor but I still think like a poor person). Social programs – we think we don't deserve them, like we're not worthy.' She explained that Aboriginal people gave up a great deal to the European settlers; the land and other sources of wealth that they relinquished should have brought plenty to everyone but did not. She employed similar terms to explain that, for whatever reasons, including personal inadequacies or weaknesses, poverty was largely beyond the control of the poor themselves. She was a truly charitable person in her estimate of others. In return, she asked of others – and this was a matter of central importance in her life – that they not judge her by standards alien to her own experience. In reacting to these redistributions of wealth by the state, she need not have been apologetic. Everyone benefited from her productivity. And, yes, although she sometimes worried that her success was temporary, she had made it over the border into the middle-income camp.

Roseanne was speaking after Canada, in company with other nations in the North Atlantic world, had adopted a public policy of privatization, deregulation, and increased reliance on the market. Canada's Free Trade Agreement with the United States (FTA), negotiated by the governments of Brian Mulroney and Ronald Reagan, which went into effect in 1989, and the North American Free Trade Agreement (NAFTA), negotiated by the governments of Mulroney, George Bush, and Mexican President Carlos Salinas in 1992–3, contributed to the reconstruction of the Canadian economy and subsequent changes in social policy, including the unravelling of parts of the social-security safety net. As Roseanne noted, it was debatable whether the programs that rescued her in the 1980s continued to operate effectively a decade later.

Given the changes in work and income experienced by Roseanne and Frank and the rest of Canada's 'plain' people in the past half-century, the thesis that new dimensions of time and space created a new economic environment seems irrefutable. Roseanne had exchanged a two-room shack for a suburban home with all the trimmings – dishwasher, thermostat, television, computer, and, best of all, plenty of room for 'personal space.' In contrast with the workplaces of Phyllis and Ali Knight, Frank's office now insisted upon endless, high-paced, and covertly monitored decision making. For both Frank and Roseanne, the monetization of time and the compression of space went hand in hand with the development of a new version of everyday private space in which surveillance had become an issue and welfare state provisions were crucial in many aspects of their daily lives. Could they reach new levels of productivity? Could they extract greater effort from mind and body? Could they expect the state's social net to rescue them, or their dependants, in time of need? Could they, at the least, maintain an atmosphere of security in their households? The answers, however tentative, require that one move away from the economic approach and into an analysis based on cultural forces and political action.

8

Culture and Politics Today

Perceptions of the dimensions of time and space that developed in the last half of the twentieth century introduced new cultural and political considerations for ordinary Canadians. The most striking was citizens' need to come to terms with such global phenomena as television and computer-based communication technologies. A second was that they began to doubt their governments and even to doubt their own effectiveness as citizens. Third, they started to question whether any nation, including Canada, was a viable, appropriate governing unit. Yet if democracy was to have real meaning, should ordinary folks not be able to exercise some control over the direction and pace of change? These were the challenges of the screen-capitalist age.[1]

I

In his portrait of Frank, Ken Dryden noted that the family spent much of its free evening time in front of the television, watching situation comedies and similar American-origin dramas of the *Simpsons* and *Cheers* and *Ally McBeal* variety: 'They have two TVs, but they are a family. What is the point in being home if three of them watch in the basement, two in the bedroom? They are apart enough, at school, with friends, listening to their own music in their own headsets or rooms. Watching TV together might not seem much, but week after week, sharing the same shows and episodes, they build reference points, they share the same experiences.' Why do they spend so much time with television, as much as the Canadian average of twenty-two to twenty-four hours a week? 'It is the poor man's night out. It is habit. But, mostly, it is a break. They don't want anything else to do. He doesn't want to be engaged, engrossed or stirred to action. He's not looking for information ... He is no weak, addicted member of the "entertain me" generation, he is tired.'

The power of these television habits cannot be denied. Nor can the complexity of this activity in which we all to some degree engage. It is no easy matter to decide what is going on, culturally and politically, when people adopt this new (compared to Phyllis Knight's daily habits) pattern of life, especially in the case of Canadian viewers who accept so much American programming as part of their daily fare.

If Canadians are watching U.S. products, they are also engaging in a process with noteworthy international, metropolitan, and cosmopolitan qualities. Roseanne's reaction to the death of Diana, Princess of Wales, illustrated the apparently homogenizing quality of a made-by-television society. Because they were grief-stricken together, Roseanne and millions of other ordinary citizens around the world might be seen as evidence that common people had lost the creative political power that their predecessors had exercised so effectively.

Certainly the 'Diana effect,' as it was called, seemed to know no boundaries and to capture the attention especially of common people. During her years of fame, Diana Spencer was depicted in the media as both a princess and an 'ordinary person,' a celebrity of surpassing beauty and wealth but also a vulnerable individual who had less formal education and had been 'unfairly treated by fate.'[2] Diana was willing, said one commentator, to be open to public scrutiny of her private emotions, to take risks in front of a world-wide audience, to play intriguingly with the customary dividing lines between public and private spheres. Moreover, she offered

an 'immediate sympathy with ordinary people.' As she herself said, 'I'm closer to people at the lower end of the scale.'[3]
In retrospect, it might seem impossible for one who belonged
among the 'supersonic rich,' who was described as 'probably
the most famous woman, perhaps the most famous person,
of her time,' to claim close connections with the common
people, but the declaration was unsurprising in the days
after her death.[4]

Roseanne responded to Diana's sudden death in 1997 with
a sustained sense of grief, an expression that seemed comparable to that of many citizens around the world. She con-
fessed that she was surprised by the depth of her reaction:

> [Diana's death] was a big thing. I don't know why, but that
> really affected me. It was a really, really depressing time ... [She
> wonders whether her reaction stemmed in part from a per-
> sonal crisis that coincided with the week of the funeral.] The
> whole world, it seemed like, was mourning, and I watched it ...
>
> It almost felt like you knew her, because she was so, such a
> public figure, and even when you listened to her interviews,
> you really could make a connection with her. My kids didn't
> feel like this – they laughed at me when I told them.
>
> When the funeral was on, and Elton John sang, that was so
> so sad. And you know, all my friends that were my age, I asked
> them, 'Did you guys watch it?' and they all did and they said
> that they bawled through the whole thing. It didn't hit me that
> hard, but it was emotional, and I was sad. My inlaws, they all
> set their alarms, and at 3:00 am they all got up to watch it.

It would be easy, especially in retrospect, to dismiss Roseanne's grief as one of many shallow expressions induced by
a television-befuddled society. It would also be wrong. Roseanne is not a shallow dupe, as you would readily recognize if
you met and talked to her, nor is Frank, nor are the ordinary
citizens of any modern society. They may make mistakes in
their assessment of public and personal situations, but they
recognize their responsibility for the analysis they arrive at

and for the choices they make. They try to be active citizens. At this juncture, when inherited institutions seem increasingly fallible and the demands of a new age are urgent, we should consider whether Roseanne, Frank, and their peers have, first, established meaningful cultural institutions and, second, used them to exercise a degree of political control. The last two sections of this chapter factor in the Quebec experience, as traced in a recent memoir, and then the global and North American contexts of instant telecommunication in which we all now live.

II

We can examine the cultural context of ordinary people's daily lives from two perspectives – the *production* and the *consumption* of culture. Under the first heading, 'production,' one might inquire whether Canadians have created a quality and range of cultural opportunities sufficient for the conduct of a satisfying national conversation. Under 'consumption,' one might ask whether, given the undoubted variety of cultural activities in which they engage, Canadians do conduct a satisfying investigation of meaning and political choice.

How should we interpret the television-watching of Frank and Roseanne? The negotiation of meaning in Frank's basement TV room represented a meeting point of private and public time and space. Frank's negotiation took place, as Benedict Anderson wrote about literacy, within the lair of his own skull, but its content was provided by public media. In these private-public moments, Frank was a producer as well as a consumer of culture.

Cultural consumption, in other words, is a fundamental part of a history of people's expression. In recent decades, common people in the rich regions of the globe have been inundated with consumer goods. For the first time in history, they have had access to a cornucopia of wonders, from 'muscle cars' and foreign holidays to music videos, televised sports, and fashionable clothes.[5] However, they did not necessarily take from it the precise messages that sponsor or creator had expected. Students of audience response (or 'reception analysis') have even argued that the reader (or consumer), not the creator, constructed the text.[6] The conjuncture of the two circumstances – ready availability of a wide variety of cultural goods at a price that common people can afford and consumers' power to elicit their own meaning from a cultural product – make interpretation of an ordinary citizen's participation in these cultural moments very difficult.[7]

Cultural power ranges from the fleeting to the profound. Pierre Bourdieu, the French student of sociology, distinguishes between 'short-term cultural commerce ... in items of limited symbolic value, and longer-term operations in which major symbolic value is dependent on the slow building of *authority*. At the level of major philosophical, literary and cultural systems, and indeed at a deeper level, cultural work, the privileged institutions – now not only universities but academies, national cultural institutions, public cultural systems – can be seen as indispensable instruments of production of the ideas and practices of an authoritative order.'[8]

Is this true? Are the privileged institutions really that powerful? One wonders about the relation between the discussion conducted in business-funded 'think tanks' and university seminar rooms, on the one hand, and the conversations of ordinary citizens, on the other. Where do common people actually discover and debate questions that matter profoundly to them? Do the patterns of analysis, as Bourdieu implied, filter down from elite institutions through the popular media to ordinary folks, thereby moulding the political reactions and cultural perspectives of the citizenry? These are questions about the consumption and production of culture. I suggest that we can answer them properly only if we keep in the forefront of our thinking the daily life and thought of people like Roseanne and Frank.

Dating the commencement of screen-capitalism and consumption culture, the era that shaped Frank and Roseanne, is crucial for Canadian cultural studies. Although this history has yet to be written, it is reasonable to suggest that the Canadian common people, as opposed to the prosperous middle class and the elites, caught only glimpses of this world in the first half of the twentieth century. Mail-order catalogues and big-city department stores and daily newspapers were the first of the communication giants that made possible the eventual supremacy of this common public culture. The cinema attained mass media status in Canada during the 1920s, and by 1936 Canadians were averaging monthly visits, twelve visits per year, to movie houses.[9] Radio sets were in one in three homes by 1931, and three of four by 1940, so this was the decade when radio too became a mass medium.[10] Magazines were increasingly popular in the 1920s and 1930s, and books became more affordable after Penguin's paperback revolution in the mid-1930s.[11] However, in a generation when unemployment and low wages and agricultural drought made incomes uncertain, ordinary Canadian households could not participate more than sporadically in the new material abundance.

By the late 1940s, a common public culture based on popular consumption – on the media and on modern design, on modern franchising, and on the automobile and modern urban planning – awaited ordinary Canadians. Then came the steady increases in household wealth between the late 1940s and the early 1970s. Significantly, the culture of consumption and the world of screen-capitalism arrived on the wings of the most accessible vehicle of communication yet devised – television – which began in Canada in the early 1950s and within the decade had spread to over 80 per cent of Canadian homes.[12] Thus Frank and Roseanne, children of the 1950s and 1960s, belonged to the first Canadian generation of common people who grew up in the screen-capitalist age.

Canada's experience with the production of culture, as opposed to consumption, was mixed, as the examples of film and radio reveal. Movies first came to Canada from the United States and were welcome foreign diversions to

Depression-era audiences.[13] With the exception of a few
uninspiring feature-length productions, Canadians did not
imitate the Hollywood example by creating a national industry for the production of entertainment films. In 1947,
Ottawa accepted an extraordinary deal by which Hollywood
studios agreed to insert mentions of Canada in their scripts
in exchange for free rein over their private property (that is,
movie screens situated in Canada) and a government pledge
not to impose quotas on showings of foreign films. One
important 'site' of the community's 'conversation' – the
movie theatre – was thus not generally available to Canadian
producers of films.

It was often pointed out that the Canadian response in
film was the creation of an alternative cultural product – the
documentary, produced by a government agency, the
National Film Board, or NFB (ONF in French) – later supplemented by the television wings of the publicly funded Canadian Broadcasting Corporation (CBC) and the French-
language Radio-Canada. There was merit in this argument.
The NFB in the 1940s and 1950s produced documentaries on
war and agriculture and social differences and then distributed the films by means of a network of projectionists who
used community halls and church basements and school
classrooms.[14] It later established regional production houses
and also dedicated one department to the film work of
women. By the 1990s the NFB had created an extraordinary
resource of nine thousand films on Canadian life. Although
the artists, organizers, and audiences built a rich cinematic
tradition, they had not challenged Hollywood's dominance
in the commercial movie palaces. Rather they created alternative sites for 'conversation' conducted within the film
medium. The relative influence of the two sites, the movie
theatre versus the schoolroom or church basement, has never
been assessed.

Radio differed slightly in that Canada built a dual system
of national public (the CBC and Radio-Canada) and local private broadcasters. Whereas Canadian radio sets could in the

early 1930s receive numerous U.S. programs but few Canadian productions from outside the region in which they were located, by the 1940s a national network conveyed a great deal of inter-regional material and constituted a significant new force for trans-Canadian and pan-francophone understanding. The influential CBC and Canadian Press (CP) news tradition was taken from the precise and professional British model, while entertainment programs borrowed from the American example and, especially in the case of private stations, relied heavily on U.S. studios and eventually on the American music recording industry. Canadian radio in sum differed to a degree from Canadian film because it adapted more successfully to the overwhelming U.S. presence. But it, too, carved out alternative sites for Canadian conversations.[15] Thus, in the story of Canada's production of cultural opportunities, the mixed results in film and radio can stand as exemplars of a general pattern: domestic producers won a few skirmishes, but Americans dominated because they set the rules of war.

The new technologies of the mass media, especially film and television, were central to the picture of the world that Frank and Roseanne put together. Roseanne recalled watching cowboy movies as a child and remembered that even the Aboriginal children from the nearby reserve refused to identify with the Indians on the screen, insisting instead on being cowboy heroes. She remembered her uncles listening to Saturday night hockey games on the radio, honouring loyalties to favourite teams and players. She can still sing the refrains of television advertisements from the 1960s when, in their house in the city, a rented TV set was on from the moment they returned home after school until well after they went to bed. Something profound must have been at work as these information sources lodged a view of the world in her mind; much of the material was American, and most of it encouraged habits of consumption.

Canadians often did not appreciate the complexity of their

responses to artistic and cultural work. The historian Donald Creighton, like many observers who wrote between the 1930s and the 1960s, accepted a quite literal, empirical reading of the relation between cultural product and consumer. What the author wrote the consumer read; what the author meant the consumer absorbed. Moreover, Creighton and other Canadian critics of 'mass culture' linked the presumed corrosive effects of an expanding consumer market to their worries about citizen gullibility and concluded that an independent Canada was doomed. His conceptualization offered no room for the Canadian people to reinterpret the American message.[16] The Franks and the Roseannes would simply be manipulated by the mass media and the gurus of the advertising and film industries. Writing in the early 1970s, when he was about seventy, Creighton despaired. What he saw as 'native Canadian moral standards and cultural value ... [and] the inherited Canadian belief in an ordered and peaceful society and a gentle and simple way of life' were being eroded:

> The truth is that Canadians, in accepting American capital, American management and American technology, have unwittingly received the spirit that animates all three. They have not only become subjects of the American continental empire; but they have also become converts to its characteristic philosophy of life, to what might legitimately be called its religion. The central doctrine of this religion is the belief that progress is the only good in life ... In the satisfaction of man's wants lies the only meaning of his existence; and human wants can, of course, be altered, multiplied and endlessly refined by modern advertising. There must be no limitation on the indefinite expansion of man's desires, no limitation on the capacity of modern industry to satisfy them.[17]

The dyspepsia sounds familiar, but does it represent Canadian realities accurately? Much cultural production occurred in the United States, was shipped into Canada, and was received by the Canadian people with gratitude.[18] Frank and Roseanne illustrated this condition. The historian Paul Rutherford has argued that, because Canadian audiences were increasingly indistinguishable from the American, the typical response in Canada has been to favour American over domestic cultural products and to take similar or identical messages from them. Similar views have been expressed in other parts of the world about domestic culture. However, these are sweeping judgments, based on inadequate evidence and simplistic models of cultural consumption, and we must not leave them unchallenged.

A cultural history of Canadian people's consumption has only begun to appear. The Canadian comedian Martin Short, commenting about his (and his many compatriots') rise to fame in American show business, said that he had studied the craft from the outside while growing up in Hamilton, Ontario: 'When Americans watch television,' he said, 'they watch television. When Canadians watch television, they watch American television.' The implications of this distancing process – an intellectual and cultural feat of no mean proportions if it really exists – have yet to be established. What we need is a careful analysis of representative episodes of Canadian cultural consumption. Only then could we decide which observer, Rutherford or Short, has the stronger case.

Canadian cultural production similarly has not been studied in terms of the degree of its completeness as a 'conversation forum' for all citizens. What is clear, as I noted in earlier chapters, is that Canadians conducted effective conversations in every type of community – labour, farm, middle class – before the Second World War. After 1945, however, as the new communication technologies became more insistent, Canadians began to worry about whether they were adapting effectively.

Concern about cultural policy led to the appointment in 1949 of the Royal Commission on National Development in the Arts, Letters and Sciences (Massey Commission). Louis St

Laurent's Liberal government asked Vincent Massey and his commission to review not only broadcasting policy but also aid to universities, government scientific research, financial assistance to students, film and museum and archival organizations, and assistance to all the fine arts, from music to handicrafts. The government sought to consider the media as only one part of the broader cultural expression of Canadians. By leaving open the possibility of continued government intervention, Canada was willing to consider the British and European models of cultural policy, in contrast to the American, wherein private foundations and the market played a greater role.[19]

The Massey Commission has been unfairly painted as having delivered merely the elitist voice of High Culture. True, the commissioners – a diplomat, three professors, and an engineer – distrusted commercialism and wished to promote a 'made-in-Canada culture' that might 'inoculate the Canadian people against the perils of vulgarization à la America,' as Paul Rutherford charged.[20] However, the commission's report also laid or rebuilt the foundations of many of modern Canada's cultural institutions. These included CBC radio and television, the Canada Council, the National Library, the National Museums of Canada, the National Film Board, and the Public (now National) Archives. Each institution had both a francophone and an anglophone presence or evolved within a bilingual mandate. None was simply a vehicle of high culture; all might be used to create messages at any level of cultural expression.

Though aware of the criticism that they represented an elite, the commissioners denied that they wanted to educate the public in matters of good taste. They claimed to be seeking to broaden the opportunity for and capacity of Canadians to 'enjoy work of genuine merit in all fields.' Their goal was to investigate 'that part of education which enriches the mind and refines the taste.' They were bound to examine how the nation set about 'the development of the intelligence through the arts, letters and sciences.' They did not shrink from describing their duty as an assessment of the nation's 'spiritual resources.' In other words, the commission was concerned less with the form of the expression than it was with its degree of excellence.

The commission's second, and higher priority, lay with the community itself. This was a matter not of high or elite culture but of the moral or spiritual qualities of the community as community: 'Canadian achievement in every field depends mainly on the quality of the Canadian mind and spirit. This quality is determined by what Canadians think, and think about; by the books they read, the pictures they see and the programmes they hear. These things, whether we call them arts and letters or use other words to describe them, we believe to lie at the roots of our life as a nation.'[21] As Northrop Frye observed several decades later, the Massey Commission represented a landmark intervention of government into the sphere of the market. Its report, he wrote, 'signified the end of cultural *laissez faire* and assumed that the country itself had a responsibility for fostering its own culture.'[22]

It is no mere whim to root the story of contemporary politics in the Massey Commission. For the next fifty years, Canadians fought and refought the battles waged at its hearings. Whatever the medium of expression – radio, television, magazines, book publishing, music, film – the issues of Canadian artists' access to domestic markets, restrictions on foreign dumping of excess production at subsidized prices, national quotas, promotional support to ensure that Canadian producers received at least some attention, and appropriate spurs to excellence dominated the debate.[23] The existence of such policy debates could not conceal one extraordinary trend: a formerly colonial community that had once produced and circulated only a little work in the established forms of cultural expression – literature, music, film, scholarship, theatre, the plastic arts – had in a few decades witnessed a flowering in these areas. The emergence of Canadian cultural heroes might seem small consolation, given the overwhelming power of Hollywood and the

American television networks. However, the presence of such stars can also be taken as evidence that Canadians, like nationals everywhere, found means to converse among themselves as well as to develop their own talented performers. There is a tautology in this experience: a viable, politically involved community requires communication vehicles, and when the dominant communication technology changes it must adapt to the new media if it is to remain a viable community.[24] Frank and Roseanne probably recognized that their community was not often reflected in the television shows that they watched. Yet they found ways to participate in the national conversation and to reflect in Canadian ways on American fare. The degree of their involvement, and the quality of their comprehension, constitute questions of cultural politics.

III

Given the necessity of community conversations, how could the people ensure that information was freely available and that debate was freely entered into? Were the political elections on which the entire system rested a valid indicator of their choices?

Frank did not often discuss formal politics. The morning after the Ontario provincial election in September 1990, he announced to the family at the breakfast table: 'In case anybody says something, the NDP [New Democratic Party] won. Just so you don't embarrass yourselves.' As his little warning suggested, the family spent almost no time on formal public affairs. Dryden recorded: 'He voted Liberal because his father voted Liberal and because the Liberals were in power and as the devil he knew, offered fewer surprises. But they lost ... He hadn't paid much attention to the campaign, just enough to say something at work if he had to. There was nothing much about it that had anything to do with him. So why be interested? Long ago, he had learned never to imagine he had in his hands what he doesn't. That was the perfect recipe for unhappiness. So political parties go their way, and he goes his. And now he knows he can live without them, even if it isn't in them to imagine he can.'

Roseanne and Frank shared a common outlook on their community and future. They believed that they should look after themselves, do what they could to ensure the security of their families, make whatever further contribution was within their power through their work, and leave the big public questions to others. Roseanne said that wealthy people ran the world and that 'the more money you have, the more power you have.' 'The poor people,' by contrast, 'are always getting the shaft.' She believed that politicians 'need to take care of themselves, and they need to say that they represent someone,' but she did not expect them to intervene with solutions to crucial problems in her life: 'The representatives of the people aren't very close to the people. And they can help you out if it's just a small item, but when it really is something that counts and can really make a difference, I think a lot of these people's hands are tied.' Her scepticism should not be interpreted as alienation from the community. In a thousand ways, big and small, Roseanne and Frank did participate in public affairs.

Although they seemed diffident about public matters, Frank and Roseanne held strong views on any number of issues. One that ranked high on their list, though for different reasons, was race. Frank, who lived among hundreds of thousands of immigrants in Canada's largest city, was concerned that his community had changed dramatically just in his lifetime. He sometimes resented the new arrivals. However, he did not like his own reactions. He would say, grudgingly and resentfully, that the newcomers should accept Canada's rules of behaviour. He had trouble with the differences that he saw and heard in stores and on the streets. One of his frequent sayings, 'I feel like a stranger in my own country,' illustrated his unease both with himself and with the changes around him.

Race was one of the most profound forces in Roseanne's

life. Her smaller city was changing because of Aboriginal migration from reserves, as well as the steady influx of new-comers from overseas, and the degree of racial difference on city streets was nearly as great as in Frank's Toronto. She was very conscious of skin colour – her own and that of other people – and she believed that other Canadians placed those who are not white in an inferior category. A fundamental aspect of her identity originated in the fact that her mother is partly Aboriginal: 'Aboriginal people get the worst of the racism,' Roseanne said, 'worse than the Asian and African Canadians.' Although she could 'pass for white,' as she put it, others in her family could not. She would never forsake them. She had chosen to place her Aboriginal inheritance (which she might easily have ignored or denied, as her mother had once preferred) at the centre of her life. Yet Roseanne also insisted that everyone in the world was fundamentally the same and wanted similar things for their families: 'We all value those good things like kindness, caring, honesty ... we all want to be happy ... and that's the way I like to look at the world.'[25]

Both Roseanne's profound commitment to the Aboriginal in her heritage and Frank's unease about immigration con-tradicted their surface responses to politics and politicians. They did know a great deal about public issues: they could actually feel the flash points in their communities. Race was just one of the matters of importance to which they reacted. A longer list might have included jobs and income, family and education, health and safety, and concerns for the com-munity's future. Roseanne and Frank did not despair about the world in which they lived. Roseanne was very positive about the future: 'I'm really excited about the future for my children. It makes me feel like ... I've done an okay job, it makes me feel like I've accomplished something, if nothing else, with my kids.' Frank would have felt the same way. What was the point of dwelling on the negative when life would be going on in any case? As long as he got up in the morning and his family needed his presence, he would be happy to deal with his small corner of the world.[26]

Frank and Roseanne might claim to ignore politics, but they did know who determined the bus fare, who ran the school, who was in charge of garbage pick-up and safety in the streets, who built the highways, who funded the hospital, and who decided foreign policy. This was not a bad list for any civics class. It suggested that ordinary Canadians, for all their professed distance from government, did participate in politics all the time. They knew who took the taxes from their paycheques and roughly on what those sums were spent.

How did they gather information, evaluate it, and make political choices? The daily lives of plain people took place within household, workplace, and a variety of community institutions. Both Frank and Roseanne, as it happened, had grown up within church-going families. Roseanne still worshipped regularly. She explained that the clergy had often helped her: 'When I've had a lot of unanswered questions, or things I've had difficulty with, I'd go to them and they would explain it to me – these are real-life things ... and it seemed to me they always made me understand it in a way that made me feel good ... It's personal, but also prayer – spiritual – I feel that's what pulled me through, all along, to have that faith.'

Frank no longer went to church every Sunday but his outlook too was shaped by views that he acquired early in life. According to Dryden, 'Faith happens to him. It is everything that isn't him and can't be him. It has to do with all the things he doesn't understand and can't control and never will; with questions which have no answers or none he will ever find. Things happen; no need to look too hard, to wonder, examine or doubt, there *is* a reason ... With faith he has put himself into hands that know and control and will watch out for him; into hands he trusts.' The traditional and textual perceptions of this world and the next, time-honoured explanations of the mysteries of life and death, whether mediated by a church or not, continued to shape the world-view of many ordinary citizens in Canada.

The family was the central institution in which meaning was established and debated. Relations between spouses, and adults' relations with children, provided forums within which to wage struggles about outlook and behaviour and to reach fundamental conclusions. Ken Dryden's portrait of Frank devotes long chapters to his grandparents and parents and to life within the family. In recounting her life, Roseanne similarly started with her father and mother and their village, noting briefly where her grandparents fitted in, and then concentrated on the world that she knew as a child. Both engaged in continuing discussions with their children and their spouses. The workplace, including casual conversations with workmates, also shaped their cultural life. It offered a meaningful structure for their daily existence, including specific time frames, task lists, and, most important of all, a slowly evolving collective outlook – one shared by all the workers – on the meaning of the work itself. The many institutions of the community also provided them with opportunities to compare opinions and swap facts. Frank and Roseanne were 'producing culture' – that is, participating in exchanges that helped them understand the meaning of their work and their lives – in all of these venues. They were steadfast in their loyalties. They sorted through television's messages, debated their meaning with family and friends, and remained true to values that were older than last year's favourite program and rooted more securely than any brand-name preferences.[27]

Somehow Roseanne and Frank found a way to cope with the flood of messages that came their way. They added them up into more or less coherent patterns and made political choices based on this analysis. A recent American study of what people knew about the world and how they formed their opinions concluded that television was central to their information-gathering, although they had other sources, too. These citizens employed just a few frameworks to capture and process the information that they gathered. Perhaps surprisingly, given the prominence of ideology in academic discussion, these frames of analysis did not necessarily fall into the standard left-right, socialist-conservative loyalties. Five main narrative frames established the plot outline – economic assessment (profit and loss, prosperity and poverty), conflict (win and lose, whether in local race relations or international affairs), power/powerlessness (notably when the weak are overwhelmed by a superior force), individual emotion (arising from concern for self and others), and morality (as in traditional teachings about doing the right thing). According to the authors of this extensive study, such frames 'actively filter, sort, and reorganize information in personally meaningful ways in the process of constructing an understanding of public issues.'[28] What the authors took from their extensive surveys was that news dissemination, despite its chaotic appearance, did provoke an active, analytical response among citizens.

Beneath this level of narrative frames, a similar British research project suggested there lay another set of frames, derived from an individual's values.[29] Thus common Canadians did not wallow in a swamp of unwanted and unheeded information as they consumed the cultural offerings available to them. As in their response to television drama, so in their perception of politics, they actively sorted and filed the material that enabled them to come to terms with the larger community.

IV

What about Quebec? The ordinary citizens whose stories I discussed above spent little time reflecting on the boundaries of Canada. They recognized that this nation provided the political context within which they lived. Quebec was different; to a lesser extent, the circumstances of French-speaking citizens outside Quebec also differed from that of other Canadians. The reasons for such a major divergence lie in history, in law, and in the facts of population totals. The

French inhabited a space sufficiently unlike that of other Canadians that it demands separate discussion.

Simonne Monet-Chartrand's writings can serve as an example of this francophone distinctiveness. Unlike the other citizens who have served as representatives of a generation and epoch, Simonne came from a privileged background. Both her grandfather and father were judges in Quebec, and both had been elected to public office. Born in 1919, she grew up among the leading French citizens of Montreal. Despite the efforts of her parents and their clerical allies, at the age of twenty-two she married a captivating critic of the status quo, Michel Chartrand and, despite a brief estrangement from her parents, embarked on a life of active political involvement and heavy family responsibilities. As she neared the end of these busy years, during the 1980s, she prepared a richly detailed four-volume autobiography that offers remarkable insights into the relatively small and integrated society of twentieth-century Montreal's French elites.[30]

The first volume actually opens on 16 October 1970. And therein lies an illustration of the distinctiveness of her society. The date will be instantly recognizable to citizens of Quebec: it is the date of the proclamation by Canada's federal government of the War Measures Act, which led to suspension of the accustomed civil liberties of all Canadians on the grounds of an 'apprehended insurrection' launched by the Front de Libération du Québec. One of the several hundred citizens arrested during this night of extraordinary police activity was Michel Chartrand, Simonne's husband. She was left alone with two of their children, 'shocked, thrown into confusion by the unexpected and rude visit by the police, their searches, their excessive powers, their arrogance.' Upset and bewildered, she then embarked on a day-long pilgrimage to the home of her grandfather, at St-Jean d'Iberville, and her parents' nearby summer home at Beloeil, where she had vacationed as a child. She felt tormented, nearly out of control. She had believed herself to be the partner of a union militant, one who talked too much, garrulously and extravagantly, but whose instincts were always for good. Now she was the wife of a political prisoner.

Returning home, she dug out trunks of family treasures, leafed through clippings and pamphlets – 'a precious patrimony' – and reached a kind of peace: 'I had rediscovered myself as the living heir of a line of men and women of character, proud and independent-minded: political dissidents.' Her choice of opening sequence for her memoir was a declaration that her adult life had been one of struggle, of political and social involvement, as well as of hard, unending labour on behalf of her family. In the twelve hundred pages that followed, she acknowledged many times that her husband was on the road more often than not defending workers and leading strikes, and yet she insisted that their love sufficed to inspire hope and drive away despair. The social and public causes, she declared, were worth the private sacrifices.

Simonne had started her adult life as an activist in the Catholic youth movement. There she encountered Michel, fresh from the Trappist monastery at Oka, and found his religious and nationalist convictions and intensity of purpose irresistible. Like so many of their contemporaries, they perceived the world through the lenses of French literature, Roman Catholic philosophy, and Québécois popular culture. Optimistic, idealistic, militant, they absorbed Mozart and Piaf, LaFontaine and Gélinas, Maritain and Hugo. They recoiled from Canada's declaration of war against Germany and opposed conscription for military service. In these intense political moments, Simonne met the galaxy of able young men and women who occupied positions of leadership in Quebec between the 1950s and the 1980s, and as her letters reveal she came to see 'French-Canadians' as having interests different from other Canadians.

When their first baby was baptized in 1943, Michel wrapped the infant in a blue shawl decorated with white lilies: 'My daughter will know into what race she was born,' he said. Simonne recorded with dismay the imprisonment in 1940 under the War Measures Act of Montreal mayor Camilien Houde for opposition to manpower registration, but she also noted that two of her brothers-in-law were officers in the Canadian army and that Houde was re-elected mayor in December 1944. In her memoir she quoted at length the first speech by her friend André Laurendeau in Quebec's Legislative Assembly – an event she attended with pride – wherein the leader of the Bloc Populaire Canadien pledged to 'defend and restore the sovereignty of Quebec in areas of its own competence within Confederation. Quebec, in my view ... is a true State, a people, a nation. We can and ought to seize the means to endow ourselves, as a majority, with laws suitable to our interests and our provincial ideals, not acting for the British Empire, the trusts, or the petty interests of the old parties.' It was a decade of intense excitement, and Simonne lived it to the full.

In the 1950s and 1960s, Michel became ever more deeply involved in union activity. As he wrote to Simonne during one prolonged absence from home, 'I truly believe that democracy and the Christian spirit will be saved by the organized working class.' Simonne was preoccupied by the needs of their seven children, but she continued to participate in community groups. She spoke to women's and church assemblies and found a new role as broadcaster and writer. She and Michel both joined the Co-operative Commonwealth Federation (CCF, or 'social democratic party,' as Michel preferred) but failed to win any but 'moral victories,' in Simonne's words. The peace movement and nationalism became more prominent in her life in the 1960s, although her speaking and writing about family matters still occupied first place in her public interventions. She became a campaigner for nuclear disarmament and travelled in 1963 to an international peace congress in Moscow – a voyage that prompted allegations that she was a secret Communist.

The events following the October crisis of 1970 ruptured the pattern of their lives. Michel, then the president of a Montreal union council, was interned without means of communication with the outside world for three weeks and was held in jail until his acquittal on charges of seditious conspiracy four months after the FLQ crisis commenced. It is impossible to avoid the conclusion that he was singled out, along with others among the 'Montreal Five,' by such old comrades in arms as Pierre Elliott Trudeau and Gérard Pelletier, to make the point that public utterances can on certain crucial occasions be criminally irresponsible. Michel and his compatriots contended with equal force that free speech was an indivisible right. Simonne was subjected to petty and not-so-petty official harassment intended to intimidate her. But Michel and Simonne were wrenched from the nationalist battle by a family tragedy, the accidental death of a daughter, and when Simonne resumed her community involvement, it slowly took new directions. Their social circles altered by the bitterness of the October Crisis, she turned to human rights and feminist concerns, reflected on her role as grandmother, joined the Rhinoceros party, and adjusted to several illnesses and her advancing years. Her energy undiminished, she also prepared the four volumes of her memoir and several other books.

What can one conclude from her testimony about the space occupied by French-speaking citizens? Simonne was a nationalist, a Christian, a pacifist, a feminist, a democrat. Her own intense public education as a youth demanded that she participate in community affairs. Her romantic idealism bound her to Michel, an equally dedicated, energetic, and combative romantic. She placed one or another part of her complex set of allegiances in the forefront at different moments in her life. On occasion, she accepted a trans-Canadian as well as a Quebec identity. Increasingly, however, under the pressures of political alliance-making and political accidents, she and Michel chose the camp of those who believed that only a sovereign state could maintain the kind of community they desired. In the end, as she wrote, she was 'entirely Quebecoise ... especially in the hope of sovereignty.'

Perhaps one-half of the French-speaking population of Quebec agreed with them; the others continued to accept a 'two-nation' status quo.

Yet Simonne also recognized the complexity of competing loyalties and public identities associated with Canada and Quebec. During a western Canadian tour in January 1971, she found great support for the Montreal Five. She wrote the imprisoned Michel that, although she was doing well in English, she had difficulty with 'words having a different sense in each language, like the word nation.' It was an interesting observation. The world of broadcasting, of Fridolin and Gilles Vigneault, of such programs as 'Radio-Collège,' 'Fémina,' and '5D,' had raised a higher wall between French- and English-language environments in North America just as increased governmental roles made the Quebec provincial state a stronger ally of the French-language nationalists.

Thus Simonne's view of Quebec space evolved during her lifetime from a Catholic (and potentially trans-Canadian) to a French-centred community. In political terms, the transition was accompanied by the assumption, so common in the decolonizing world of the 1960s and 1970s, that national status carried with it an obligation to be 'sovereign.' As the Chartrand family's experience in October 1970 showed, the contest over two irreconcilable views of nation made for tense and at times bitter struggles within the Canadian state.

V

The nation also played an important role in the lives of Roseanne and Frank. They had grown up believing in the virtues of Canada, and, although they found plenty in it to criticize, they assumed that it was one of the unshakeable institutions in their world. Had it, too, lost relevance? Was the nation, whether Canada or Quebec, still the appropriate political unit within which to make community decisions?

At the heart of this discussion was the relative role of markets and governments in the community. This debate had reached an important moment in the last decades of the twentieth century. The problem lay in the nature of politics itself. More specifically, it rested in the communication processes that enabled citizens to choose a course of action for their communities. Electronic communication had altered the balance in the historic struggle between government and market forces for power in the community.[31]

Harold Innis perceived a half-century ago that any form of communication possesses a bias in the sense that it permits some groups to exercise greater control than others over community discussions. This bias could harden into a monopoly, he added, 'when groups came to control the form of communication and to identify their interests, priestly or political, with its capacity.'[32] Innis, like his friend Creighton, was unsympathetic to consumerism. Unfortunately, this lack of sympathy for a central aspect of the new age led him to phrase questions about cultural production unwisely. Innis warned in 1952 that 'commercialism' represented a great danger to the integrity of the Canadian border. What he meant by commercialism was the power of the market. He defined it as 'a system that ultimately transferred all control from the person and community to the price system: where people are fed every product, including knowledge, by a machine they merely tend.'[33] However, Innis's definition was problematic. He had phrased the issue boldly, but perhaps his choice of analogy – the market as a machine that the people merely tended – was too simplistic, and too slanted in favour of the market's critics, to be convincing. After all, consumers exercised choice whenever they bought or rejected a product. What Innis should have focused on was the relative power of the community and the individual, the relative worth of market-mediated choices as opposed to those choices made by the community *after* public debate and when it was consciously considering *community*, not individual priorities.

Frank and Roseanne were only vaguely aware of these debates. They knew that there were regular news stories about American magazines in Canada, American television programming, and American moviemakers' access to Canadian screens, but they did not dwell on such matters. Nevertheless, they had been participants, wittingly or not, at every stage of this contest between market and nation. Their votes, their answers to polling organizations, and their preferences in leisure time activities had helped to reshape the country.

'Freedom of the press,' a phrase enshrined in the First Amendment of the U.S. constitution, played a part in this contest. The amendment put beyond the reach of political debate, according to Harold Innis, the priorities of 'a specific historical and political setting ... [and made them] the expression of a universal law.' Innis was writing in the late 1940s and early 1950s, before television had made its mark and when radio still looked like a mere supplement to newspapers in the shaping of political life. He argued that the American doctrine assumed 'the sanctity and efficacy of market mechanisms.' The doctrine assumed too the unwisdom of government intervention, even if intended to promote such sensitive imponderables as national identities.[34] Innis believed that the allegedly universal principles in the U.S. constitution actually gave constitutional protection to one type of technology over others and in this way restricted rather than expanded freedom: 'Freedom of the press has been given constitutional guarantees as in the United States [and] has provided bulwarks for monopolies which have emphasized control over space.' By this, Innis meant that large newspaper corporations could exercise control over increasingly large units of market territory. 'Under these conditions,' he went on, 'the problem of duration or monopoly over time has been neglected, indeed obliterated.' By 'time,' Innis meant each citizen's firm sense of continuity between past and future, ensured by mechanisms of community decision making.[35]

Innis argued that the newspaper had been granted a monopoly over knowledge. This did not apply to all knowledge, of course, but to the information necessary to one's life as a citizen in a community. He believed that in the long run, through its promotion of news and advertisements, the power of the U.S. newspaper, reinforced by the First Amendment, spread the 'spatial bias' implicit in the medium. In his terms, American imperial priorities, expressed as a determination to treat every consumer equally and as an *individual* consumer, infringed on Canadians' equally urgent and valid claims to control over time, over the nation – that is, over community discussion. Innis would have said that nations at the centre of great empires had little interest in the historic forces that knitted peripheral communities together. They had every commercial reason to promote their own space-oriented bias. As a result, successive American governments of both political stripes exported a national approach to industrial strategy in the guise of a universal principle concerning freedom of cultural expression. The peripheral nation would have to suffer the consequences.

Canada as nation was being challenged by developments in communication technology, but so were other nations. A means of understanding the transition may lie in Roseanne's response to the week of mourning after the death of Diana Princess of Wales. There was something different in the worldwide response to Diana's death as compared to previous global moments. The reaction did contain an element of group-driven excess, but what was unusual was that the demand of the people guided the media, rather than the reverse. As Roseanne said, forcefully: 'I don't think we were manipulated by the media because ... people wanted that, it seemed like. People wanted to know every single detail. I felt I knew her before – I just felt she was someone I knew. I connected with a lot of things that she felt – I guess all of the turmoil she went through – that's the connection I made with her.'[36] In this sense, Roseanne was a typical citizen of the global village. She felt that she knew Diana well because of photographs, television news coverage, and magazine stories. She was certain that she had not been tricked into feeling something alien to her own being. Her reaction merits careful scrutiny. And the analysis must begin with communication technology.

In Roseanne's response, one witnessed, first, the extraordinary cultural power of the modern communication revolution: from the invention of telegraph and camera in the nineteenth century to the globe-spanning satellite connections of today, these media made possible the nearly universal recognition of – and, what was more, identification with – particular circumstances and events. This change was a product partly of simultaneity, partly of the rendering of still and of moving images in a fashion that seemed truer than life.[37]

Within the developments in communication technology there lay a second phenomenon: the emergence of celebrity or fame as a form of communication. People knew intimately the biography of a few select individuals whose behaviour in moments of testing was said to reveal fundamental truths of character and circumstance. These truths were easily grasped by hundreds of millions of observers. Thus individuals selected for media superstardom were able to create 'a receptacle for our own most complicated emotions.'[38] They had in themselves become media.

Third, the communication revolution and celebrity as media had paradoxically strengthened and weakened nations. Ernest Gellner's valuable insight into nationalism was based on the thesis that modern communication systems are necessary to the proper functioning of modern societies. Gellner suggested that nations and the national cultures spawned within them represented 'a response to the mobility and anonymity of modern societies. Far from being a movement of retreat from the modern world, nationalism is a solution to some of its most distinctive problems.' In a world where labour is so mobile, where continuous innovation is the rule, and where local economies have been subsumed within larger production systems, citizens require a means of establishing a shared vocabulary: 'Communication among people who do not know each other ... is a functional necessity.' This communication was conveyed by a 'common language' that could carry meaning 'not dependent on local contexts and usages.'[39]

Let me try to explain. For Britons, the Diana phenomenon was unleashed by the death of a celebrity, but of a celebrity who was entrenched so deeply in the history and institutions and literary patterns of a nation that it seemed like the demise of a beloved member of the immediate family. For citizens of many other countries, Diana was also a celebrity, but she fitted within fewer institutions and historic patterns – she was merely Cinderella and princess and jet-setter perhaps – and her death thus could be described as that of a familiar acquaintance. Yet in Canada, and perhaps in other parts of the Commonwealth, among watchers of a certain age, Diana's funeral week also marked the passing if not of a member of the immediate family perhaps, of someone who was more than a familiar acquaintance. The moment might be depicted from Roseanne's vantage point as the death of a close relative. The depth of her response, and the variety of reactions in a range of social groups, offered lessons in the continuing cultural variety of the modern world.

Why did Roseanne's response differ from that of Americans, for example, or Brazilians, or of anyone else in countries outside Britain? Consider the context in which Roseanne placed her story. She had actually initiated this discussion with me by describing popular culture in her childhood community. The very sequence of her conversation revealed her thinking. In response to an open-ended question about her childhood leisure time, she had spoken about the entertainment venues and the celebrities who stood out in the public life of her little town during the 1950s and 1960s: 'Elvis Presley was big then, too. I remember him in the theatre – his movies – and the Beatles were big too. We used to get those Beatle rings ... They were from England, like the Queen ... The Queen was a big thing too. We had to sing "God Save the Queen" every day. And when John F.

Kennedy was killed, that was a really big thing in the community – the school was let out early that day.'

The succession of her thoughts in this passage, especially
given that she then turned to the funeral of Diana, revealed
her particular upbringing and perspective on the world. The
juxtaposition of the American and the British, in a stream of
consciousness that seemed so natural – almost a single
breath – demonstrated the Canadian situation of her generation and era. She recognized without a moment's reflection
that the United Kingdom and the United States were powerful and that both had loomed large in her child's vision of the
world. She also knew that only the United Kingdom
belonged in the daily rites of Canada's own public sphere.
She then turned, without taking a breath, to the mourning
for Diana. And in this comment she expressed the perspective of older Canadians. She, like them, was accustomed to
the central role of 'the Royals' in public life. That the British
context and the royal family have only recently been
removed from much of the ritual of community events
reflects the reality of a new generation in Canada. Roseanne's children laughed at their mother's expression of grief.
The very sequence of Roseanne's conversation illustrated her
world-view. She was a consumer of cultural messages that
circulated around the globe; she was also a Canadian and, in
some residual way, a member of the Commonwealth; and,
finally, she was North American.

The emergence of this North American reality must be
acknowledged. The advent of new electronic communication
systems after 1950 extended the American advocacy of space-
based freedom as a principle in information management to
cross-border flows of data and data-based corporate services.
At the United Nations and other international forums,
including the Uruguay Round of the General Agreement on
Tariffs and Trade (GATT), the United States insisted that the
free flow of information was sacred and that national government interventions in such flows would be inappropriate.[40]

These principles also constituted the essence of the Canada-U.S. Free Trade Agreement (FTA). A 1989 briefing memo to President George Bush explained: 'The US sought to preserve liberalizing steps taken by the Mulroney government and prevent a return to inward-looking, nationalistic policies of the 1970s, especially in energy, investment, banking and services.'[41] Three of the four areas mentioned involved sectors that relied heavily on electronic communication.

As a result of the FTA and NAFTA, signed in 1988 and 1993, respectively, continent replaced Canada as the political space in which physical goods were traded. However, this economic continentalism was not supposed to prevail in cultural expression. Cultural goods were excluded specifically from the FTA because Canadians recognized that these goods possessed a dual character, as both economic product and community conversation. The task for the Canadian people was to understand this dualism, to come to terms with the global corporate strategy that might inhibit their conversations, and to find the means of asserting their own interests – and the interests of other common people around the world – against the inevitable excesses of an unrestricted market.

It is too early in the development of the screen-capitalist society to say whether Canadians have adapted well or badly to its communication technologies. Frank and Roseanne have found ways of speaking to each other and of expressing some of their political views. Like other Canadians, they watched Canada's English-language television news, tuned in to American television drama, and assisted their children as best they could to adapt to a changing world. Simonne participated much more intensely in community issues but increasingly cut herself off from discussions of community priorities in the rest of Canada. She placed a premium on language as the source of the community's historic distinctiveness and thus permitted the invisible walls erected by electronic communication to narrow the community within which she carried out her public responsibilities.

The children and grandchildren of these three citizens in their turn will carry on the story that we know as history. By setting up households and teaching their offspring, they will learn what is meant by participation in and responsibility for the local, the provincial, the national, the continental, and the global community. One hopes that they too will learn to distinguish among the networks of conversation in which policies are debated and choices are made and, even more important, to preserve the vitality of each of these spheres of citizenship.

Conclusion

This narrative has responded to Canada's circumstances today by creating a new version of national history. It claims that all citizens can benefit from the sense of continuity that historical synthesis alone can provide. It is political because it contributes to a community discussion about politics and the responsibilities of citizenship.

The book's first message concerns the practice of national history. I recognize that this is a time of awkwardness in Canadian historical circles. For a century, many Canadians, especially those who believed that they belonged among the leaders of society, have assumed that the country's present boundaries demarcated a logical, defensible, even superior society. Yet every nation is now under siege in a new global economy. Every imagined community is challenged by global communication systems. In Canada, the continuing campaign for Quebec sovereignty challenges the bilingual/ multicultural consensus of the preceding generation as well as the physical unity of the country. Political allegiances in other parts of the country also set 'bottom lines' that challenge the status quo. To make matters worse, there are few histories that tell the nation's story in terms sufficiently broad to unite the many factions quarrelling in our midst yet local enough to evoke the 'feel' of Canada's distinctive places.

Part of the awkwardness in Canadian history arises from the appearance of literally dozens of new subjects, and hundreds of new books, in the past thirty years. It is obvious that gender history, labour history, Aboriginal history, and all the other new perspectives in social studies are not going to go away. These works of scholarship represent important new views that must be acknowledged in the narratives that historians create. My story builds on this remarkable generation of historical innovation and tries to incorporate its lessons. This book is part of the new social and cultural history.

A different challenge to a citizen's historical reflection emanates from the apostles of poststructuralism and the postmodern. These approaches, which assert the duplicity of all language and the utter instability, even unknowability, of individuals and communities, seem to me to fly in the face of the obvious political institutions with which we do live. Yes, institutions such as the nation will change, as they have in the past, but for the moment the existence of communities in defined spaces, and the consequent indissoluble relation between community and place, an association consolidated by our political system, cannot be gainsaid. If Canada conforms very poorly to the classic European definition of a nation, it is none the less a political state, and its citizens make judgments on important political matters. To do so with confidence, Canadians must be assured that their country – their 'nation' – has a certain definable relevance.

The citizen's sense of security will come in part from historians' contributions to synthesis. This is my second concern. New histories must establish a convincing case for the existence of continuity between past and present. Some recent commentators have noted the fragmented nature of the systems of communication in our society, which deprives ordinary citizens of their ability to synthesize information. My discussion of time and space as the changing dimensions in which people make their way responds to this concern by pointing out one path to synthesis. There is, I believe, a powerful logic in the four-step historical experience of time-space dimensions in Canada. To convince citizens of that logic is, I suggest, a worthy goal.[1]

Can a history built on communication and culture such as I have written here, one that is so broad-gauged as to be nearly international, actually serve the interests of one peripheral nation rather than further the progress of a universal, homogeneous state? It is true that this country might seem indistinguishable from the United States, or Australia, or from Quebec perceived as a nation on its own, for that matter, if a discussion of time and space and communication were all that I could offer. But the illustrations in the chapters above link the theme of changing dimensions to the story of specific individuals and groups in a particular place – families fishing in the Arctic, hunting in Labrador, changing jobs in Vancouver, carrying out a variety of tasks in contemporary cities. Moreover, my discussion ties these families into the long historical evolution of society in northern North America while addressing these international themes. Therefore the synthesis is specific to this place, these people, and this nation. Canada's political circumstances will continuously change but the country has endured for well over a century and may last a good while longer. I suggest that there is comfort to be derived from the story of ordinary families' adaptations to drastic changes in the past. They have demonstrated a continuing ability to rebuild their lives in the face of significant developments in the technology of communication and in perceptions of time and space.

Third, this book represents a response to contemporary discussions about history itself. In a recent essay on historical writing, Keith Jenkins wrote that he was seeking 'a series of histories that helped us to understand the world that we live in' and then rephrased his ambition to say that he sought a series of histories of the present.[2] Just as grandmother Andre, Elizabeth Goudie, Phyllis Knight, and Frank, Simonne, and Roseanne have told stories, so this narrative adheres to a particular set of story-telling conventions. Like theirs, my narrative attempts to contribute to a reader's understanding of change and to provide assurances of continuity. As befits a volume composed today, my story acknowledges that it is only one of many ways to establish a sense of perspective. But it also tries to be transparent by acknowledging that all these statements carry 'historiographical' – that is, interpretive – implications. After all, I *selected* the stories that I have told to explain my present purposes as well as to illuminate the past. And there can be no doubt that I am wearing my heart on my sleeve when I defend the country of Canada, argue for a sense of continuity in this place from past to present, and insist that the experience of ordinary households has its own processes of adaptation and resistance.

What does my story add up to? It suggests that Aboriginal perceptions of time and space were the first to prevail in northern North America. These dimensions were based in communities that used speech as the dominant form of communication. As the story told by grandmother Andre suggests, the purpose of traditional Aboriginal story-telling, which embodies simultaneously historical and contemporary observation, is broader and less specific than that of the print-based cultures that developed in Europe. Such Aboriginal stories may seek to educate, to stake claims, or simply to establish or maintain relationships. They are open-ended, in that listeners may reach their own conclusions. These narratives comprise 'facts enmeshed in the stories of a lifetime [that] provide a number of insights about how an understanding of the past is constructed, processed, and integrated into one's life.'[3] Far from the objectivity, precision, and lessons that are the goal of Europe's literate history, the Aboriginal narrative will lay no claim to being better or worse than another account. Crucial to the interpretation that I have placed on them, Aboriginal stories emerge from particular places, families, and events. They associate empirical fact with myths as inseparable parts of a single sphere and discuss human and animal or plant life as elements that exist on the same plane as the dream world. There is an other-than-human quality to the dimensions of space in the Aboriginal world. European conventions that divide the human actor from the environment, and this world from the next, do not prevail in the traditional Aboriginal depiction of life's dimensions. The sharp contrast between the two visions constitutes powerful testimony to the distinctive character of time-space relations in traditional Aboriginal culture.[4]

Literacy and international trade in household-produced commodities accompanied the arrival of Europeans in northern North America. The dimensions of time and space that also came with the new arrivals – what may be described as the new dominant culture – eclipsed Aboriginal narrative conventions, although they could no more erase them than the advent of literacy in Europe could obliterate the oral patterns in Homeric style narratives. The common people among the European newcomers, just as much as the governors and priests and fur-trade officers, lived within the structures of thought established by written documents.[5]

The people who inhabited these 'textual communities' and whose economies relied on household production, like the Goudie family of Labrador, surveyed a limited, local world. The perceptions of time and space in their Europe-descended social order found their source in natural dimensions, just as did those of their Aboriginal neighbours. Needs determined work, the sun and the seasons were the measure of time, and the physical effort of humans or animals bridged space.[6] Yet their trade in staples ensured that the circulation of trade goods and capital from farther afield affected their 'island communities.' Their association with church, law, and government also introduced broader currents of information. In these households, the Christian calendar was one of the fundamental instruments of time division, although it existed in juxtaposition with traditional festivals and with traditional means of celebration such as the special meal, both of which resonate with Aboriginal experience. Similarly, scientific medicine was juxtaposed with folk medicine, biblical injunctions on family relations in the workplace with international trends in commodity prices, thunder gusts of local protest with laws decreed by distant courts. For Europe's offspring in northern North America, the dimensions of time and space differed from those formerly accepted by Aboriginal people of pre-contact days, but also from the dimensions being established by the leaders of Europe. Louis Riel inhabited this textual environment but was familiar with both the formerly dominant dimensions of the pre-contact Aboriginal community and the purely print-based communication order that was rapidly taking shape. Because of his simultaneous understanding of three cultural perspectives – his grasp of Aboriginal foretelling, printed European history, and the textual society between – Riel was able to win and retain the loyalty of Métis hunters.

The world's subsequent shift to nearly universal literacy and to long-distance communication media offered more diverse opportunities for comparative thought. Citizens were able 'to perceive the world around them as a particular cultural creation' and therefore to debate within that spiral of greater verbal precision that one associates with print-based societies.[7] Thus the extension of print-capitalism, the invention of the modern nation-state, and the development of the first wave of the mass media introduced another version of time-space relations to northern North America between the mid-nineteenth and the mid-twentieth centuries. The railway actually seemed to diminish distance, the newspaper to shrink time, and the political boundary to reconstruct community. Because the dominant culture redrew the dimensions in which they lived, ordinary people had to adjust to a new version of reality.[8]

In this third epoch of time-space relations, as the story of Phyllis Knight illustrated, time and space could be altered for profit. Common people in response had to adjust to new dimensions of daily life, including the rules of property and wage as well as of a national state. In making this adaptation, they added the analytic methods and communication strategies of a literate world to the lessons that they had learned from their ancestors.[9] During the first half of the twentieth century, ordinary citizens immersed themselves in the outpourings of communication vehicles, including the daily newspaper, the radio, film, and the mass circulation magazine, but also of institutions that they devised for self-defence. They learned thereby to participate in the institutions of the state, including the political party, the public school classroom, and the industrial relations system. They established their own vehicles – farm movement, women's institute, cooperative, craft union, and industrial union – to challenge those who sought to control their lives. Many of us continue to see the world in terms developed during this remarkable era and even to wonder why society does not respond as we expect when we pull the political levers that served our grandparents so well.

Time and space have since been annihilated, according to the cliché of today. The world after 1945, 1960, 1973 – take your pick – is said to be drastically different from the one before. What these exaggerations actually mean is that the dimensions of time and space have seemingly altered yet again. The passage of time has speeded up, distances have again diminished, to the point that the geographer David Harvey can describe the remaking of global space-time relations since the 1960s as introducing a tension between space and place.[10] Simonne, Frank, and Roseanne had to come to terms with this society. Their nation and neighbourhood no longer possessed the physical characteristics, no longer had the same exclusive claim as communities, that they did fifty or a hundred years ago.

In the last two decades or so, even the intellectual universe of the preceding century has exploded. Disciplines have been 'decentred,' 'metanarratives' dismissed, and seemingly venerable institutions such as the nation and the family turned into the subject of debate. As George Steiner has written, 'It is this collapse, more or less complete, more or less conscious, of those hierarchical, definitional value gradients (and can there be value without hierarchy?) which is now the major fact of our intellectual and social circumstances.'[11]

At this point, historical synthesis again becomes relevant.

Not just relevant – it becomes essential. As Eric Hobsbawm suggested in his recent history of the twentieth century, 'Under the impact of the extraordinary economic explosion of the [post-1945] Golden Age ... with its consequent social and cultural changes, the most profound revolution in society since the stone age, the branch [on which our communities sat] began to crack and break. At the end of this century it has for the first time become possible to see what a world may be like in which the past, including the past in the present, has lost its role, in which the old maps and charts which guided human beings, singly and collectively, through life no longer represent the landscape through which we move, the sea on which we sail. In which we do not know where our journey is taking us, or even ought to take us.'[12] Such gloomy but perceptive rumination requires a response in each society that possesses the political power to act.

The key to this volume is its reliance on the experience of ordinary citizens. My contention is that, in societies peripheral to the great imperial centres, the crucial measures of adaptation and community-building – the acts deserving of the term 'historic' – are those undertaken by ordinary citizens. In a complex community such as modern Canada, these accommodations are difficult to discern amid the hurly-burly of the media and the market-place. Yet I am convinced that the four great epochs of time-space dimensions remain current in Canadian society and crucial to an understanding of Canadian public consciousness. To be a Canadian today is not a matter of birthplace, race, language, ethnicity, religious affiliation, genealogy, or some combination of these characteristics. To be a Canadian is to accept certain relations with others, to adopt a specific, historically moulded vocabulary, and to work within certain institutional constraints while debating meaning within the community. To be a Canadian is to orient oneself to the past according to community choices made during the preceding centuries. To be a Canadian is to adjust to the inevitable contingency of the nation itself. To be a Canadian is a matter of circumstances that have been summarized as 'relational, cultural, historical, and contingent.'[13] This neat quartet, borrowed from sociologist Charles Tilly, who is in turn synthesizing several decades of academic work on public identity, has provided the foundation of my story.

By 'relational,' I mean to say that Canadian-ness resides in the ties *among* people, not embedded within the brain cells or bloodstreams of each individual. The hoary notions about innate nationality – Scots parsimony or Italian excitability – cannot be sustained if they are taken to be genetic inheritances rather than acquired by social interaction and education. We live in a plural world, and Canada is one of the world's great national experiments in pluralism. To imagine that each corpuscle in each citizen carries national defining characteristics is to conjure up wondrous internal tortures for those of us who descend from many heritages, but it also is nonsense. To be Canadian is to accept certain relations with one's co-citizens.

By 'cultural,' I imply that Canadian-ness consists of the way in which one experiences, explores, reproduces, and communicates meaning itself.[14] As this society has supplemented its historic trades in fish, fur, timber, and wheat with the production of machines and ideas, Canadians have learned that cultural activities generate profits even as they communicate values and skills. Citizens now debate the necessity of protecting or subsidizing certain kinds of cultural product (a book or film or athletic performance), or certain arenas of public conversation (a radio network, a dance studio, a television channel, a news-stand), because they are discovering that culture's role in community life has grown more contentious – and even fragile – as it has increased in economic significance. To be Canadian is to accept that the relative priority of the community itself constitutes an urgent aspect of this debate about values.

By 'historical,' I am insisting that at any moment a community contains and is shaped by accumulated forces that endure over much longer periods than might seem possible at first glance. Is it really imaginable that a story told round a campfire fifteen hundred years ago might offer insights into contemporary Canada? Is it conceivable that a ninety-minute battle in Quebec waged nearly two hundred and fifty years ago between a few thousand soldiers from two European countries affects events today? Or that the achievement of a political confederation of communities in northern North America in 1867 still has relevance? The continuing existence of Canada provides the answers.[15]

Public identity, as it is understood in today's scholarly discussions – that is, the negotiation among individuals of the community-wide qualities that they find favourable for such a collective enterprise as a nation – is described by Charles Tilly in the three terms ('relational,' 'cultural,' 'historical') that I have just noted, but he also adds a fourth: 'contingent.' By this he is saying that a public identity can misfire, is subject to stunning accidents, may divide into a dozen pieces or may cease to exist entirely. Canadians are of necessity conscious of the fragility of their national community.

If even the nation is divisible, then the stories that one tells about it must be the cause of fractious debates. And history is not a single, unalterable, and uncontested tale. Even the most obvious truths of one generation may seem stale platitudes or outright errors in the next. Particularly in this century, history exists in tension with national democracies. To preserve the nation in a time of so much change, leaders rally patriots with simple, clear messages that stir the majority. But what about the heartfelt loyalties of those citizens – often in minorities – who see a different truth? Lord Acton, writing in the nineteenth century, proposed that those nations most favoured in all creation accommodated many groups within their ranks. Canadians cherished the Actonian formulation during the heady, confident days of Expo 67 and the Centennial celebrations of 1967. Out of the 1960s too came an enthusiasm for Canada's 'limited identities,' for loyalties associated with the smaller and more immediate communities of class and ethnic group and region. Histories of Aboriginal people, local communities, gender, alternative sexuality, and the environment have proliferated during the last quarter-century. So much has this been the case that several historians have attributed disunity in Canada to the subsequent loss of a national narrative. These critics contend that the country lacks a history of the public community around which citizens can rally and which, in its ready appeal to patriotism, might provide an extra measure of confidence in a time of global stress. Not to mention in times of potential Quebec separation.[16]

My purpose in writing this book has been to explain in new terms why 'Canada' is a meaningful public identity. I wanted to demonstrate that the various groups of Canadians who express such a variety of responses to the world and seem to be divided by such profound barriers also share much. Because they talk together within shared community institutions and evaluate their experience in the light of a common inheritance, they also possess aspects of a common identity. Canadians often pride themselves on the diversity of their country. The great physical differences that distinguish the tundra from the prairies, the Pacific coast from the Atlantic, and the cities near mountains from those near the Great Lakes declare that the Canadian place has no single physical character. The fundamental force of linguistic difference between English and French, and also among dozens of Aboriginal and immigrant tongues, similarly separates communities one from another. Is it possible to package so many narratives, such complex processes, into a single, brief, easily repeatable lesson about what Canada is and who Canadians are? About how Canadian society coheres and does not cohere? These are the challenges of national history and of the nation itself. If nothing is shared, the community has indeed disintegrated. I suggest that the history, the cultural debates, and the relations made vital by both do knit together this diverse collection of individuals and groups.

Another way of expressing my theme is to say that this book has addressed Canadian citizenship education. Every individual in a community is engaged in education. All of us, all the time, are learning from our daily experiences and our daily contacts, from magazines and movies and mountain streams. As one scholar put it, 'the struggle to learn, describe, understand and educate is a central and necessary part of our humanity.'[17] Such a process relies partly on specific texts, the great books that have been admitted to the canon of each country. However, my story has turned away from the famous texts in order to inquire into another source for the understanding of collective experience, the broader currents of thought that circulate around us. They in turn depend on forces that we place in the category of communication and culture – the institutions and forms in which ideas, information, and attitudes are transmitted and received. This is where Simonne, Frank, and Roseanne and the Andre, Goudie, and Knight families come in. I believe that Canadians have failed to integrate the achievements of common people into the ideas and symbols that articulate their sense of nationhood, an omission that dilutes their understanding of their own strengths and narrows their appreciation of the world. To learn about their struggles and achievements is to acquire a broader education in citizenship.

This book offers an interpretation of everyday history and everyday citizens. It contends that a crucial strength of Canada lies in its common people. It asserts the creativity of every citizen, not just the powerful few. It proposes that the country, both in its weaknesses and in its strengths, finds a priceless aspect of its character in these citizens' continuing struggles to define their own perspectives in the face of change that they cannot entirely anticipate or control. Their efforts to understand are crucial acts of citizenship that enable them to claim possession of the community and its fate. Individuals' adaptations to changes in the dominant time-space dimensions – adaptations among traditional Aboriginal cultures, among settlers in northern North America's textual societies, among national citizens shaped by print-capitalism, and among citizens today – testify to the genius of common people working together in community. They demonstrate the remarkable capacity of cultures to absorb change and to ensure continuity. They demonstrate as well the absolute necessity in any society of an active and historically aware citizenry. The Aboriginal people, the settlers, and the national citizens shaped by print-capitalism are all present in today's Canada. Their stories offer a degree of hope to those who follow the history of inter-racial and cross-cultural contact. Our task is to appreciate this complexity, to be open to its influence, and to come to terms with its possibilities.[18] Grandmother Andre, Elizabeth Goudie, Phyllis Knight, and Simonne, Roseanne, and Frank would expect nothing less.

Notes

Introduction

1 Raymond Williams, *Britain in the Sixties: Communications* (Harmondsworth: Penguin Books 1962), 12, paraphrased and adapted.

2 Ramsay Cook, 'Nation, Identity, Rights: Reflections on W.L. Morton's *Canadian Identity*,' *Journal of Canadian Studies* 29, no. 2 (summer 1994), 5–18.

3 W.L. Morton, *The Canadian Identity*, 2nd ed. (Toronto: University of Toronto Press 1972). Morton offered a number of sweeping conclusions: 'British America was left [in 1846] the country of the northern economy, the economy of the great staples of fish, fur, and timber. Habitable farm lands there were, but small in relation to the whole, enough to feed the staple trades and even to export surpluses, but not enough to support a population comparable with that so rapidly growing in the United States. Thus though a continental country, Canada was still dependent, as the northern maritime frontier had always been. Moreover, though continental in extent, it was still surprisingly maritime in character. The Atlantic islands and Acadian peninsula, the St. John, the St. Lawrence, Hudson Bay, the Saskatchewan, the Fraser, the Gulf of Georgia, it was by their sea inlets and inland waterways that British America lived.' (30).

'The northern character springs not only from geographical location, but from ancient origins in the northern and maritime frontier of Europe. Within that area from mediaeval to modern times there is discernible a frontier of European culture developing across the northern latitudes in which the forward movement was largely by sea. It was not a Turnerian frontier, but it was a frontier in every sense, and it was this frontier which began the exploitation and settlement of Canada. Many of its characteristics survive in Canada to this day, and presumably will continue to do so indefinitely.' (89).

'This, then, is the first orientation of Canadian historiography. Canadian history is not a parody of American, as Canada is not a second-rate United States, still less a United States that failed. Canadian history is rather an important chapter in a distinct and even an unique human endeavour, the civilization of the northern and arctic lands. From its deepest origins and remotest beginnings, Canadian history has been separate and distinct in America ... History is neither neat nor categorical; it defines by what is central, not by what is peripheral. And because of this separate origin in the northern frontier, economy, and approach, Canadian life to this day is marked by a northern quality ...' (93).

'Dependence is second to northern orientation: dependence economically, strategically, politically – both for markets and for the supply of the needs of the mind and body which raised life on the northern frontier above the level of subsistence.' (94).

4 R. Cole Harris, 'Power, Modernity, and Historical Geography,' *Annals of the Association of American Geographers* 81, no. 4 (1991), 680.

5 As Anthony Giddens has written, the power of the favoured groups relies 'in large part on the management of time-space relations in particular locales;' cited in ibid., 677

Chapter 1: Genealogy and Economy

1 Harold A. Innis, *The Fur Trade in Canada: An Introduction to Canadian Economic History* (New Haven: Yale University Press 1930; Toronto: University of Toronto Press 1962), 392.

2 Canada, Royal Commission on Aboriginal Peoples, *Report*, no. 1, *Looking Forward, Looking Back* (Ottawa 1996), 5.

3 Until recently, European Canada's awareness of Aboriginal history has been shallow. Not every observer has been guilty of patronizing the First Nations, but a cultural blindness can be traced from the earliest commentators in the seventeenth century to the writers of history textbooks in the decades after the Second World War. When addressing Aboriginal issues, this perspective was race-based, and, intentionally or not, it often misrepresented and demeaned the diverse Aboriginal heritages. Writing about the Canada that he encountered in 1970, Robin Fisher, a Canadian historian of New

Zealand origin, said that it was 'astounding how little most Canadians knew about the cultures and history of the country's indigenous people.' Robin Fisher, 'Historical Writing on Native People in Canada,' *History and Social Science Teacher* 17 (1982), 72.

What Canadians did know was even more shocking. In an oft-cited 1971 survey of eighty-eight books that appeared most frequently on Canadian university history course lists, James Walker reported that Aboriginal people were commonly referred to as 'savages' and worse. Often Indians just 'materialized' to play a role, whether as ally or enemy, in a story told from the point of view of the European newcomers. And then they disappeared again. A number of textbooks treated Aboriginal people only in an introductory chapter on geography and environment as 'part of the setting.' They were 'other,' not part of 'us,' and their cultures were expected to wither away in the presence of a superior European 'Manifest Destiny.' Walker notes that Aboriginal people were described as being variously 'cruel,' 'treacherous,' 'bloodthirsty,' 'dirty,' 'cowardly,' 'lazy,' 'barbaric,' 'fiendish,' 'credulous,' 'grotesque,' 'superstitious,' 'gluttonous,' and 'fickle.' In these various texts, they were alleged to have behaved like thieves, drunkards, and butchers. James W. St G. Walker, 'The Indian in Canadian Historical Writing,' Canadian Historical Association *Historical Papers* (1971), 22–7.

Research and writing since 1970, as well as a significant shift in public attitudes to cultural difference, have drastically revised this portrayal. The questions that scholars ask today about the Aboriginal heritage reflect greater public awareness of Aboriginal standards of judgment and patterns of living.

4 Graydon McCrea, *Summer of the Loucheux: Portrait of a Northern Indian Family,* available from Tamarack Films, 11032 – 76 St, T5B 2C6, Edmonton, Alberta, phone (780) 477-7958. The film, 27 minutes and 45 seconds long, was shot in 1979, and released in 1983. I would like to thank Graydon McCrea for several helpful conversations and for a video copy of his film. He also made arrangements so that I could talk about the contents of this chapter with Alestine Andre, who was in 1999 the executive director of the Gwich'in Social and Cultural Institute in Tsiigehtchic, Northwest Territories (formerly Arctic Red River). I am very grateful to Alestine for giving this story her detailed attention and for sharing her understanding of the film and the culture discussed herein. I would like to

thank Joe MacDonald of the National Film Board, Winnipeg, who made these contacts possible.

5 Olive Patricia Dickason, *Canada's First Nations: A History of Founding Peoples from Earliest Times* (Toronto: McClelland and Stewart 1992), 27.

6 More than a half-million Canadians acknowledge 'some' Aboriginal ancestry, according to census takers, but only 135,000 claimed 'Metis' status in the census of 1991.

7 Frank Howard, 'Native Policy May Be Based on Faulty Census Numbers,' Ottawa *Citizen*, 1 April 1993 (I would like to thank Ed Dahl for this reference).

8 Surveys include Arthur J. Ray, *I Have Lived Here since the World Began: An Illustrated History of Canada's Native People* (Toronto: Lester Publishing, Key Porter Books, 1996); Dickason, *Canada's First Nations*; J.R. Miller, *Skyscrapers Hide the Heavens: A History of Indian–White Relations in Canada* (Toronto: University of Toronto Press 1989); Bruce G.Trigger, *Natives and Newcomers: Canada's 'Heroic Age' Reconsidered* (Montreal: McGill-Queen's University Press 1985); Robin Fisher, *Contact and Conflict: Indian–European Relations in British Columbia, 1774–1890* (Vancouver: University of British Columbia Press 1977); Leslie F.S. Upton, *Micmacs and Colonists: Indian–White Relations in the Maritimes, 1713–1867* (Vancouver: University of British Columbia Press 1979). For purposes of clarity, most of the following narrative focuses on the hunting bands, but the generalizations are intended to apply more broadly to the Canadian Aboriginal heritage.

9 Arthur J. Ray, *Indians in the Fur Trade: Their Role as Hunters, Trappers and Middlemen in the Lands Southwest of Hudson Bay 1660–1870* (Toronto: University of Toronto Press 1974).

10 Movement was also to be expected when natural disasters (such as forest and prairie fires) or epidemics struck. Assiniboines, for example, saddened and weakened by hundreds of deaths attributable to the arrival of European disease in their camps in 1781–2, left behind the land that they had occupied and moved west to reconstruct their communities. Ray, *Indians in the Fur Trade*, 104–7.

11 Mary Black-Rogers, 'Varieties of "Starving": Semantics and Survival in the Sub-Arctic Fur Trade, 1750–1850,' *Ethnohistory* 33 (1986), 353–83.

12 Marshall D. Sahlins, *Stone Age Economics* (Chicago: Aldine-Atherton 1972).

13 Adrian Tanner, *Bringing Home Animals: Religious Ideology and Mode of Production of the Mistassini Cree Hunters* (St John's: Institute of Social and Economic Research, Memorial University of Newfoundland, 1979), 56.

14 A striking illustration is the story told in Edwin James, ed., *A Narrative of the Captivity and Adventures of John Tanner during Thirty Years Residence among the Indians in the Interior of North America* (1830; reprint, Minneapolis: Ross & Haines 1956).

15 Hugh Brody estimates that 59 per cent of the livelihood of three of the eleven bands in the BC district that he studied in the 1970s (1,000 people belonged to the eleven bands in 1978) came in the form of meat from hunts, 31 per cent from furs and handicrafts, 8 per cent from wages and 2 per cent from social payments. Hugh Brody, *Maps and Dreams: Indians and the British Columbia Frontier* (Harmondsworth: Penguin Books 1981), 190–213; also Morris Zaslow, *The Northward Expansion of Canada 1914–1967* (Toronto: McClelland and Stewart 1988), 271–2; another work on hunting is Richard K. Nelson, *Hunters of the Northern Forest: Designs for Survival among the Alaskan Kutchin*, 2nd ed. (Chicago: University of Chicago Press 1986).

16 A helpful volume, despite its questionable central thesis, is Calvin Martin's *Keepers of the Game: Indian–Animal Relationships and the Fur Trade* (Berkeley: University of California Press 1978); it can be read in conjunction with Shepard Krech, III, ed., *Indians, Animals, and the Fur Trade: A Critique of Keepers of the Game* (Athens: University of Georgia Press 1981).

17 Brody, *Maps and Dreams*, 186–7.

18 James, ed., *A Narrative*, 45–9, 164–5; Tanner did not believe – or so he recorded in this narrative – that the woman had actually made this discovery in a dream. He attributed the band's good fortune to her careful tracking of the animal. See his comments on 52, 108, 140, 180–5, and 252.

19 Jennifer S.H. Brown and Robert Brightman, *'The Orders of the Dreamed': George Nelson on Cree and Northern Ojibwa Religion and Myth, 1823* (Winnipeg: University of Manitoba Press 1988), 60–2 and 8–9.

20 Thomas Flanagan, ed., *The Collected Writings of Louis Riel/Les Écrits Complets de Louis Riel*, vol. III (Edmonton: University of Alberta Press 1985), 531–2, 'Address to the Jury [Regina] 31 July 1885.'

21 Tanner, *Bringing Home Animals*.

22 One discussion of this theme is Eric R. Wolf, *Europe and the People without History* (Berkeley: University of California Press 1982).

23 Trigger, *Natives and Newcomers*, especially 183–94; Robin Fisher, *Contact and Conflict*; Daniel Francis and Toby Morantz, *Partners in Furs: A History of the Fur Trade in Eastern James Bay 1600–1870* (Montreal: McGill-Queen's University Press 1983); Charles A. Bishop, *The Northern Ojibwa and the Fur Trade: An Historical and Ecological Study* (Toronto: Holt Rinehart and Winston 1974); Paul C. Thistle, *Indian–European Trade Relations in the Lower Saskatchewan River Region to 1840* (Winnipeg: University of Manitoba Press 1986).

24 Arthur J. Ray, 'Periodic Shortages, Native Welfare, and the Hudson's Bay Company 1670–1930,' in Shepard Krech, III, ed., *The Subarctic Fur Trade: Native Social and Economic Adaptations* (Vancouver: University of British Columbia Press 1984), 16; also Ray, *The Canadian Fur Trade in the Industrial Age* (Toronto: University of Toronto Press 1990).

25 One illustration of this generalization is Frank Tough, *'As Their Natural Resources Fail': Native Peoples and the Economic History of Northern Manitoba, 1870–1930* (Vancouver: University of British Columbia Press 1996).

26 Peter Douglas Elias, *The Dakota of the Canadian Northwest: Lessons for Survival* (Winnipeg: University of Manitoba Press 1988). Government policy is discussed in Sarah Carter, *Lost Harvests: Prairie Indian Reserve Farmers and Government Policy* (Montreal: McGill-Queen's University Press 1990). Many Canadians will be surprised to see the contrast between the large Aboriginal reservations in the western United States and the small reserves of prairie Canada. The map comparing these reserves is displayed in *People to People, Nation to Nation: Highlights from the Report of the Royal Commission on Aboriginal Peoples* (Ottawa 1996) 33, and is adapted from Robert White-Harvey, 'Reservation Geography and Restoration of Native Self-Government,' *Dalhousie Law Journal* 17, no. 2 (fall 1994), 588.

27 Marshall Sahlins, 'Cosmologies of Capitalism: The Trans-Pacific Sector of "The World-System,"' *Proceedings of the British Academy* 74 (1988), 8, 9.

Chapter 2: Interpreting Aboriginal Cultures

1 Graydon McCrea, *Summer of the Loucheux* (1983).

2 Note the unusual juxtaposition of terms – can feelings actually have

a structure? – that Raymond Williams employs to explain the elusive concept of culture. Williams first developed this discussion in *The Long Revolution* (Harmondsworth: Penguin Books, 1965), 64, and 57–88, and returned to it in *Culture* (Glasgow: Fontana Paperbacks, William Collins Sons, 1981).

3 Such a vision may have to be challenged today, but I leave that discussion to part IV of the book.

4 Anthony F. Aveni, *Empires of Time: Calendars, Clocks, and Cultures* (New York: Kodansha International 1995), 81.

5 Ibid., 183.

6 Ibid., 209.

7 CBC Information Radio, Noon Edition, transcript of interview script for discussion conducted by Jim Rae, 'Smith Keeper Saves Man,' 6 Jan. 1995, Winnipeg CBW. I would like to thank Jim Rae for digging through his files to recover this manuscript. The rescuer, who speaks Ojibwa, did not feel comfortable about being interviewed in English and left this task to his truck-driving friend.

8 John L. Steckley, 'Why I Study an Extinct Canadian Language,' *Language and Society* 37 (Jan. 1992), 11–12.

9 Hugh A. Dempsey, *Crowfoot: Chief of the Blackfeet* (Edmonton: Hurtig 1972), 14, and *A Blackfoot Winter Count* Occasional Paper No. 1 (Calgary: Glenbow-Alberta Institute, 1965).

10 Aveni, *Empires of Time*, 122.

11 Claude Lévi-Strauss, *The Savage Mind* (Chicago: University of Chicago Press 1966), 262–3; the object of such traditional peoples, he wrote, was 'to grasp the world as both a synchronic and a diachronic totality.'

12 Calvin Martin, ed., *The American Indian and the Problem of History* (New York: Oxford University Press 1987), 16.

13 Robin Ridington, *Trail to Heaven: Knowledge and Narrative in a Northern Native Community* (Vancouver: Douglas and McIntyre 1988), 57–8.

14 This version is quoted from Robin Ridington, 'Fox and Chickadee,' in Martin, *American Indian*, 132; it appears in different form in *Trail to Heaven*, 70–1.

15 Martin, 'Epilogue: Time and the American Indian,' in *American Indian*, 207–8.

16 Ridington, 'Fox and Chickadee,' 134.

17 Catherine McLellan et al., *Part of the Land, Part of the Water: A History of the Yukon Indians* (Vancouver: Douglas & McIntyre 1987), offers

insights into the stories and cultures of the Dene. For other cultures, see, for example, James R. Stevens, *Sacred Legends of the Sandy Lake Cree* (Toronto: McClelland and Stewart 1971); Basil H. Johnston, *Ojibway Heritage* (Toronto: McClelland and Stewart 1976) and *Tales the Elders Told: Ojibway Legends* (Toronto: Royal Ontario Museum 1981); Charles Clay, *Swampy Cree Legends* (Toronto: Macmillan 1938; Bewdley, Ont.: Pine Ridge Publications 1978); Brown and Brightman, *'The Orders of the Dreamed,' George Nelson on Cree and Northern Ojibwa Religion and Myth, 1823* (Winnipeg: University of Manitoba Press 1988); a remarkable illustration of the transfer of actual events into mythical narrative is D. Wayne Moodie, A.J.W. Catchpole, and Kerry Abel, 'Northern Athapaskan Oral Traditions and the White River Volcano,' *Ethnohistory* 39 no. 2 (spring 1992), 148–71; a sample of interpretive approaches is provided in Jennifer S.H. Brown and Elizabeth Vibert, eds., *Reading beyond Words: Contexts for Native History* (Peterborough: Broadview Press 1996).

18 G. Malcolm Lewis, 'Indian Maps: Their Place in the History of Plains Cartography,' *Great Plains Quarterly* 4, no. 2 (spring 1984), 104.

19 Hugh Brody, *Maps and Dreams: Indians and the British Columbia Frontier* (Harmondsworth: Penguin Books 1981), 266–7. Not all works of craft and "art" carry the same interpretive freight. Quillwork, the decoration of clothing with coloured porcupine quills, is a distinctive Aboriginal practice, particularly among certain hunting bands. These decorative materials on the clothing of northern Indians impressed early travellers such as Alexander Mackenzie. When he encountered Dogrib and Slave Indians in the 1789 expedition on the river that now bears his name, he reported: 'Their clothing is made of the dressed skins of the rein or moose deer, though more commonly of the former ... Some of them are decorated with an embroidery of very neat workmanship with porcupine quills and hair of the moose, coloured red, black, yellow and white ... Their cinctures and garters are formed of porcupine quills woven with sinews, in a style of peculiar skill and neatness.' Cited in Kate C. Duncan, *Northern Athapaskan Art: A Beadwork Tradition* (Vancouver: Douglas & McIntyre 1989), 33. Duncan argues that Athapaskans readily adopted European materials and designs precisely because this work did not normally have sacred or symbolic meaning. The rapid adoption of European styles illustrated their exceptional ability to borrow from other cultures in order better to meet the challenges of

a very harsh environment: 'The Athapaskan openness to new ideas, a trait developed in the battle for survival, inclined them to whole-heartedly embrace a myriad of cultural introductions, among them floral bead embroidery. They integrated the art into their cultural system and in short order developed a dynamic range of variations. Today, bead embroidery continues to play an important role in Northern Athapaskan culture' (189–90).

20 This statement by Robert Bear appears in the exhibit on Aboriginal elders that was on display in the Prince Albert Museum in 1996. I would like to thank Tom Ferris and Bill Smiley for arranging this visit.

21 James Axtell, *The Invasion Within: The Contest of Cultures in Colonial North America* (New York: Oxford University Press 1985), 14–5; also 'The Power of Print in the Eastern Woodlands,' *William and Mary Quarterly* 3rd Series 44 (April 1987), 300–9; note the work of Walter Ong, *Orality and Literacy: The Technologizing of the Word* (London: Routledge 1982), and of Jack Goody, *The Logic of Writing and the Organization of Society* (Cambridge: Cambridge University Press 1986) and *The Interface between the Written and the Oral* (Cambridge: Cambridge University Press 1987).

Rosalind Thomas writes: 'One cannot produce hard and fast rules, but what has emerged from the vast and sophisticated anthropological literature on oral tradition is that oral traditions, dependent on their human transmitters and on human memory, do not get passed on accidentally: the reasons for remembering them and passing them on are fundamental. [People] select the tradition and may well change it in the process: the reasons for change can be cultural, social, political, or ideological. Similar effects have been analysed by psychologists working on human memory.' *Literacy and Orality in Ancient Greece* (Cambridge: Cambridge University Press 1992), 109.

22 Axtell, 'The Power of Print,' 304–5, and *The Invasion Within*, 102–3.

23 Axtell, 'The Power of Print,' 306–7.

24 Evidence of such continuity can be seen in such influential publica-tions as Bruce G. Trigger, *Natives and Newcomers: Canada's 'Heroic Age' Reconsidered* (Montreal: McGill-Queen's University Press 1985); Abraham Rotstein, 'Trade and Politics: An Institutional Approach,' *Western Canadian Journal of Anthropology* 3, no. 1 (1972), 1–28; John S. Milloy, *The Plains Cree: Trade, Diplomacy and War, 1790 to 1870* (Win-nipeg: University of Manitoba Press 1988).

25 Marshall Sahlins, 'Cosmologies of Capitalism: The Trans-Pacific Sector of "The World-System,"' *Proceedings of the British Academy* 74 (1988), 7.
26 Brody, *Maps and Dreams*, 36–7; Janet Chute, *The Legacy of Shingwaukonse: A Century of Native Leadership* (Toronto: University of Toronto Press 1998).
27 John S. Milloy, 'The Early Indian Acts: Developmental Strategy and Constitutional Change' in Ian Getty and A.S. Lussier, eds., *As Long As the Sun Shines and Water Flows: A Reader in Canadian Native Studies* (Vancouver: University of British Columbia Press 1983); A.C. Hamilton and C.M. Sinclair, *Report of the Aboriginal Justice Inquiry of Manitoba*, vol. 1, *The Justice System and Aboriginal People* (Winnipeg: Queen's Printer 1991), 62–75; Katherine Pettipas, '*Severing the Ties That Bind': Government Repression of Indigenous Religious Ceremonies on the Prairies* (Winnipeg: University of Manitoba Press 1994); Douglas Cole and Ira Chaikin, *An Iron Hand upon the People: The Law against the Potlatch on the Northwest Coast* (Vancouver: Douglas and McIntyre 1990).
28 Pettipas, *Government Repression*; Miller, *Skyscrapers Hide the Heavens*, 206–7.
29 Carter, *Lost Harvests*; Jean Friesen, 'Position Papers for Treaty One Interpretation' and 'Treaty Five Interpretations' (unpublished, Treaty and Aboriginal Rights Research, Winnipeg), 1985.
30 John L.Tobias, 'Canada's Subjugation of the Plains Cree, 1879–1885,' *Canadian Historical Review* 64, no. 4 (1983), 519–48.
31 Flanagan, *The Collected Writings of Louis Riel*, vol. III (Edmonton: University of Alberta Press 1985), 'Pétition à "votre excellence en conseil"' (7?[sic] Sept. 1884) 27–9.
32 T.F. McIlwraith, University of Toronto, to Joint Senate–House of Commons Committee on the Indian Act, 12 May 1947, No. 21, 1537.
33 Mary Lou Fox, 'Speaking the Language of the Soul,' *Globe and Mail*, 15 April 1993, A16.
34 Barry Came, 'Fighting for the Land,' *Maclean's* (27 Feb. 1995), 16.
35 Harold A. Innis, *The Fur Trade in Canada: An Introduction to Canadian Economic History* (New Haven, Conn.: Yale University Press 1930; Toronto: University of Toronto Press 1962), 391–2.

Chapter 3: Elizabeth Goudie and Canadian Historical Writing

1 The memoir is recorded in pencil in school scribblers in Elizabeth

Goudie, 'A Labrador Trapper's Wife,' ms., n.d. (c. 1963), Newfoundland Centre, Memorial University of Newfoundland. The edition cited in these excerpts is Elizabeth Goudie, *Woman of Labrador*,
ed. David Zimmerly (Toronto: Peter Martin Associates 1973; Halifax: Nimbus Publishing Company 1996). I am grateful to Dan Soucoup of Nimbus for permitting me to cite this edition. A sample of
Elizabeth's original punctuation is in Gerald M. Sider, *Culture and
Class in Anthropology and History* (Cambridge: Cambridge University Press 1986), 158–61. I have chosen to follow Zimmerly's edition
because this is the version that Elizabeth herself approved.

By an extraordinary coincidence, I was able to contact Elizabeth's
son, Joe Goudie, a former cabinet minister in Newfoundland, and
to hear a little about the family, including the news that Horace had
published his own memoirs, *Trails to Remember* (St John's: Jesperson
Press 1991). The coincidence was the appearance at my house of
Bernie Howgate selling his books door to door. I am indebted to
Howgate, author of *Journey through Labrador*, who encouraged me to
telephone Joe Goudie.

2 The double wedding took place at the Grenfell Mission and
included a supper and a dance that lasted until 3:00 in the morning,
at which her new brother-in-law played the violin: 'He had had TB
and his right arm was cut off about six inches below his elbow.
They tied the bow onto the stump of his arm and he played for our
wedding dance. He could play the violin well.'

3 David Zimmerly, 'Introduction' to Goudie, *Woman of Labrador*, xviii.

4 In 1673 the residents of New France numbered about 7,000 peasants, clerics, and soldiers. In 1773, still occupying a narrow strip of
land along the St Lawrence River and outposts as far-flung as
Acadia, the western interior, and Louisiana, their descendants numbered about 70,000. They were joined in the next half-century by
approximately 50,000 'Loyalists,' who stayed with the British crown
by moving north after the American Revolution, and by perhaps
another 100,000 people of British origin who settled along the
Atlantic coast, in Upper Canada, and in the fur trade districts of the
'North-West' during the first decades of the nineteenth century.

This segment of the British Empire, which grew steadily as a
result of emigration from the British Isles, consisted of eight regions
of population and economic activity scattered across the northern
half of the continent. In the west, by 1860, the districts around Victoria and the lower BC mainland had attracted 10,000 people to work

in fur and gold and farming enterprises. In the Red River Valley of
Manitoba 7,000 people, four-fifths of mixed Aboriginal and European origin, worked on farms and in the fur trade. Britain's four
Atlantic colonies were larger – Prince Edward Island (80,000), Newfoundland (124,000), New Brunswick (250,000), and Nova Scotia
(330,000), for a total of about 800,000 in 1860. They supported some
agriculture and fishing but also, especially in New Brunswick, a
great deal of shipbuilding and lumbering.

The largest concentrations of population in British North America were near the St Lawrence River and the lower Great Lakes; by
1860, 1.1 million people lived in Canada East (Quebec) and 1.4 million in Canada West (Ontario), over 80 per cent on farms and in
small towns. Here was the heartland of a potential nation – the richest soil, the growing manufactories, the focus of political activity,
and the dominant voices (French and English) in intellectual life.
Outside these eight regional communities lived Aboriginal hunting
families and a few isolated agents of commerce – by the rich fisher-
ies along the Pacific coast, in the forests of the North-West near Fort
Simpson and Edmonton House and The Pas, at the subarctic whal-
ing and fur stations such as Fort Churchill and York Factory, and
near the fur posts of northern Ontario and Quebec and Labrador.

Modern Canada has its origins in the Confederation in 1867 of
the United Province of Canada, New Brunswick, and Nova Scotia,
and in the subsequent annexation of Manitoba and the North-West
(1870) and the admission of British Columbia (1871) and Prince
Edward Island (1873). At that time, its population was about one-
tenth that of its southern neighbour, the United States, and of its
French and British 'parents.' Thus by the mid-nineteenth century,
Canada belonged in the company of small European settlements
overseas – Peru, Argentina, and Venezuela in the Americas, Australia and New Zealand in the Pacific.

In the twentieth century, two considerable waves of immigration,
from 1900 to 1913 and during the 1920s, expanded the influence of
European cultures in Canada and added a small number of Asians
and Africans to the national population. This pattern resumed after
1945. By the end of the 1960s, Canada had adopted a more 'colour
blind' immigration policy, and since 1970 it has admitted several
hundred thousand newcomers each year. The total population rose
from five million in 1900 to fourteen million in 1951 and just over
thirty million at the end of the century.

5 Elliott Merrick, *True North* (New York: Charles Scribner's Son 1933), 7, 8, cited in David William Zimmerly, *Cain's Land Revisited: Culture Change in Central Labrador, 1775–1972* (St John's: Memorial University of Newfoundland Institute of Social and Economic Research 1975), 208–9; also Goudie, *Woman*, ix–x.

6 I do not want to paint a picture that is too rosy. Certainly, as Elizabeth's editor noted, white trappers ran into conflicts with Indians over trap lines. Zimmerly, *Cain's Land Revisited*.

7 The Métis were descendants of mixed European and Aboriginal marriages. Their communities – French-speaking on the Atlantic and Great Lakes, English-speaking on Hudson Bay, and Russian-speaking on the Northwest coast – developed over a span of three centuries. Some of these families established a distinctive and tenacious group identity, especially in parts of the prairies and the north. These Métis acted as the bridge between communities, the translators and guides and freighters who acquired a distinctive third 'culture' or 'identity' in the process of serving two other communities. Defined more by language and occupation than by 'race,' their separate status was eventually reflected in their politics. One rarely discussed consequence of the 1982 patriation of Canada's constitution has been the official recognition of a distinct status for these self-identified people. Section 35 of the Constitution Act, 1982, explicitly acknowledges the 'existing aboriginal and treaty rights' of not only the Indian and Inuit but also the 'Métis peoples of Canada.' For an introduction to this community, see Jennifer S.H. Brown and Jacqueline Peterson, eds., *The New Peoples: Being and Becoming Métis in North America* (Winnipeg: University of Manitoba Press 1985).

8 Northrop Frye, 'Conclusion to a *Literary History of Canada*,' in Northrop Frye, *The Bush Garden: Essays on the Canadian Imagination* (Toronto: Anansi 1971), 217, 220, 225; the essay was first published in 1965. Frye read this literature, and the pioneer experience represented therein, as more complex than any single-hued version might suggest. He saw the positive and the pastoral or comic in both literature and the public past.

9 W.L. Morton, *The Canadian Identity*, 2nd ed. (Toronto: University of Toronto Press 1972), 113–14.

10 Morton, *The Canadian Identity*, 4. Both parts of his sentence deserve scrutiny: by 'the greatest of civilizations,' Morton would have meant the British parliamentary tradition and the Judaeo-Christian

religious inheritance; and by his reference to the grimmest of environments, this Manitoba historian implied that the effort necessary
to survive in such a climate and to maintain political control over
such a vast and difficult terrain was nothing short of heroic.

11 Margaret Atwood, *Survival: A Thematic Guide to Canadian Literature*
(Toronto: Anansi 1972), 31–5, 54. Atwood's literary guidebook proposed that 'survival,' an extension of Frye's discussion of garrison
mentality, was the 'unifying and informing' symbol of Canadian literature. Atwood emphasized the challenges posed by natural forces
to the pioneers: 'For early explorers and settlers,' she wrote, life in
Canada 'meant survival in the face of hostile elements and/or
natives; carving out a place and a way of keeping alive.' The hostil-
ity, lest there be any doubt, emanated from 'the land, the climate
and so forth.' Susanna Moodie, an English gentlewoman and writer
in pioneer Canada, encountered obstacles in 'swamps, bogs, tree
roots and other immigrants,' according to Atwood. In later generations, Atwood acknowledged, the obstacles to human existence
were more numerous and more likely to lie within the psyche than
in the natural environment. None the less, she reminded her readers, 'Death by Nature ... is an event of startling frequency in Canadian literature.'

The alienation attributed to imaginative expressions of earlier
centuries was not simply a matter of human fear. As Frye commented, 'I have spoken of what I call a garrison mentality and of
the alternating moods of pastoral optimism and imaginative terror
(which has nothing to do with a poet's feeling terrified) in earlier
Canadian writing.' But the impression of a people besieged in a
hostile land was now rooted in cultural interpretations and in Canada's selective tradition.

12 The economists W.A. Mackintosh and Harold Innis are especially
associated with the Laurentian, or staple thesis. Mackintosh rel-
ished the paradox that 'Canada is a nation created in defiance of
geography, and yet the geographic and economic factors have had a
large place in shaping her history.' Innis, in his survey of the fur
trade, concluded that 'The northern half of North America
remained British because of the importance of fur as a staple product.' On Mackintosh, see 'Economic Factors in Canadian History'
(first published 1923) in Carl Berger, ed., *Approaches to Canadian History* (Toronto: University of Toronto Press 1967), 1–14; on Innis, see
his *The Fur Trade in Canada: An Introduction to Canadian Economic*

History (New Haven, Conn.: Yale University Press 1930; Toronto: University of Toronto Press 1962).

13 Allan Greer, *Peasant, Lord, and Merchant: Rural Society in Three Quebec Parishes 1740–1840* (Toronto: University of Toronto Press 1985), 66–7; W.J. Eccles, *France in America* (Toronto: Fitzhenry & Whiteside 1973), and *The Canadian Frontier 1534–1760* (Hinsdale, Ill.: Dryden Press 1969).

14 Greer, *Peasant, Lord, and Merchant*, 46–7. One Englishman complained of temperatures of 73°F in the 'stuffy' houses! However, despite the early use of homemade goods, A.B. McCullough reports that as much as one-half of the cloth consumed in rural households in the Quebec City area in early nineteenth century was imported; McCullough, The Primary Textile Industry in Canada: History and Heritage, ms. report, Parks Canada, 1992.

15 R.M. McInnis, 'Marketable Surpluses in Ontario Farming, 1860,' in Douglas McCalla, ed., *Perspectives on Canadian Economic History* (Toronto: Copp Clark Pitman 1987), 55; also McInnis, 'Perspectives on Ontario Agriculture, 1815–1930,' *Canadian Papers in Rural History*, vol. VIII (Gananoque: Langdale Press 1992), 17–127.

16 Sam Ateah, *Memories of Victoria Beach* (Victoria Beach: n.p. 1991), 50.

17 Rusty Bittermann, Robert A. Mackinnon, and Graeme Wynn, 'Of Inequality and Interdependence in the Nova Scotian Countryside, 1850–70,' *Canadian Historical Review* 74, no. 1 (1993), 1–43. Donald Creighton wrote of the 'hardy, self-reliant simplicity' of this life, the three authors note.

18 The frontier approach depicted the farm district as an egalitarian, democratic society. All pioneers embarked with a more or less equal chance, or so the assumption went, and grit or hard work or sheer cussedness would select the winners. Because the continent possessed abundant resources, wastefulness might be expected, but all who gave an honest effort could expect to reap rewards. This interpretation is customarily associated with the name of the American historian who first advanced the 'frontier hypothesis,' Frederick Jackson Turner. Implicit rather than explicit in early discussions of the frontier was the presumption that the economy would be organized according to the conventions of the emerging laissez-faire, capitalist economic system, the structure of which had already been determined in Europe. Competitiveness, markets, survival of the fittest, specialization, and comparative advantage characterized this version of the New World. The state would not

control the pioneer farm economy; the individual and the free market would.

The existence of this same distribution of farm wealth in Canada West as well as in Nova Scotia challenges the frontier hypothesis. Labourers in Canada West cleared one to five acres per year on average, but not surprisingly some farms grew larger and more prosperous, while others languished. Half of the farm households before 1840 had less than thirty acres cleared, whereas fewer than 5 per cent possessed one hundred acres or more in cultivation. Peter Russell, 'Upper Canada: A Poor Man's Country? Some Statistical Evidence,' *Canadian Papers in Rural History*, vol. III (Gananoque: Langdale Press 1982), 136–8. In one county, Peel, one of four rural households in 1835 was either renting or squatting on land. David Gagan, *Hopeful Travellers: Families, Land, and Social Change in Mid-Victorian Peel County, Canada West* (Toronto: University of Toronto Press 1981), 34. For the next generation, farmers and their sons and daughters travelled from the back country to work as labourers and domestic servants in the longer-settled townships in order to sustain the family's frontier farm. Joy Parr, 'Hired Men: Ontario Agricultural Wage Labour in Historical Perspective,' *Labour/Le Travail* 15 (1985), 94–5; John McCallum, *Unequal Beginnings: Agriculture and Economic Development in Quebec and Ontario until 1870* (Toronto: University of Toronto Press 1981).

For an introduction to this discussion, one might commence with E.J. Hobsbawm, 'Scottish Reformers of the Eighteenth Century and Capitalist Agriculture,' in Hobsbawm et al., eds., *Peasants in History: Essays in Honour of Daniel Thorner* (Calcutta: Oxford University Press 1980), 3–26; Allan Kulikoff, 'The Transition to Capitalism in Rural America,' *William and Mary Quarterly* 46 (1989), 120–44, and *The Agrarian Origins of American Capitalism* (Charlottesville: University Press of Virginia 1992); James Henretta, 'Families and Farms: Mentalité in Pre-Industrial America,' *William and Mary Quarterly* 35 (1978), 3–32; Steven Hahn and Jonathan Prude, eds., *The Countryside in the Age of Capitalist Transformation: Essays in the Social History of Rural America* (Chapel Hill: University of North Carolina Press 1985). A Canadian version is discussed by Carl Berger, *The Writing of Canadian History: Aspects of English-Canadian Historical Writing 1900–1970* (Toronto: Oxford University Press 1976), 118–22.

19 W.L. Morton, 'Agriculture in the Red River Colony' (1949), reprinted in A.B. McKillop, ed., *Contexts of Canada's Past: Selected*

Essays of W.L. Morton (Toronto: Macmillan 1980), 69–86; Gerhard
Ens, *Homeland to Hinterland: The Changing Worlds of the Red River
Metis in the Nineteenth Century* (Toronto: University of Toronto Press
1996).

20 Innis's discussion of the 'staple trap' – a community's inflexible reliance on one staple product – has relevance to the Métis experience.
Gerhard Ens's volume on Métis agriculture and the hunt takes this
tack because it hinges in part on the near-extinction of the buffalo in
the 1870s. After four to seven generations of fruitful adaptation to
the land, and one generation devoted to capitalist speculation, the
prairie Métis entered a period of crisis. Markets for a local product,
merchants eager to exploit the trade, and new technology in the
form of guns and railways had doomed an entire community to
severe dislocation. The limits of the resource had been exceeded.

21 Martine Segalen, *Love and Power in the Peasant Family: Rural France in
the Nineteenth Century* (Chicago: University of Chicago Press 1983),
9; see, however, Marjorie Griffin Cohen, *Women's Work, Markets and
the Economy in Nineteenth-Century Ontario* (Toronto: University of
Toronto Press 1988), and, for another dissenting opinion, Judith M.
Bennett, *Women in the Medieval English Countryside: Gender and
Household in Brigstock before the Plague* (New York: Oxford University Press 1987) and '"History that stands still": Women's Work in
the European Past,' *Feminist Studies* 14, no. 2 (1988), 269–83.

22 An interesting illustration of this view is presented in Katherine
Martens and Heidi Harms, *In Her Own Voice: Childbirth Stories from
Mennonite Women* (Winnipeg: University of Manitoba Press 1997),
'Maria Reimer,' 37–8.

23 Morris Altman, 'Economic Growth in Canada, 1695–1739: Estimates
and Analysis,' *William and Mary Quarterly* 45, no. 4 (Oct. 1988), 684–
711. I would like to thank Shane White for this reference.

24 R. Cole Harris, 'The Extension of France into Rural Canada,' in
James R. Gibson, ed., *European Settlement and Development in North
America: Essays on Geographical Change in Honour and Memory of
Andrew Hill Clark* (Toronto: University of Toronto Press 1978), 43–4,
cited in Roberta Hamilton, 'Feudal Society and Colonization: A Cri-
tique and Reinterpretation of the Historiography of New France,'
Canadian Papers in Rural History, vol. VI (Gananoque: Langdale
Press 1988), 98.

25 Hamilton, 'Feudal Society,' 95–110; Allan Greer estimates that feudal dues took 'a substantial portion, probably more than half,' of

the Lower Richelieu peasants' agricultural surpluses; *Peasant, Lord and Merchant,* 137.

26 Greer, *Peasant, Lord, and Merchant,* 184–93; Fernand Ouellet, *Economy, Class, and Nation in Quebec: Interpretive Essays* (Toronto: Copp Clark Pitman 1991); Gérard Bouchard, *Quelques Arpents d'Amérique: Population, économie, famille au Saguenay 1838–1971* (Montreal: Boreal 1996); R. Cole Harris, 'Of Poverty and Helplessness in Petite-Nation,' *Canadian Historical Review* 52, no. 1 (March 1971), 23–50; Allan Greer, 'Wage Labour and the Transition to Capitalism: A Critique of Pentland,' *Labour/Le Travail* 15 (1985), 7–24; Serge Courville and Normand Séguin, 'Rural Life in Nineteenth-Century Quebec,' Canadian Historical Association Booklet 47 (Ottawa 1989); McCallum, *Unequal Beginnings.*

27 An earlier statement of this view is Mackintosh, 'Economic Factors in Canadian History,' 8–14; a helpful review of the question is Douglas McCalla and Peter George, 'Measurement, Myth, and Reality: Reflections on the Economic History of Nineteenth-Century Ontario,' *Journal of Canadian Studies* 21, no. 3 (fall 1986), 71–86.

28 McInnis, 'Marketable Surpluses,' 48.

29 Were these pioneer farms populated by incipient capitalists? Were the farm families bent on the accumulation of capital, accustomed to the impersonal operation of markets, and determined to avoid the intervention of the state? This question has provoked much discussion. Recent contributions that refer to the classic work by V.C. Fowke as well as to the literature of the 1980s are Ian MacPherson and John Herd Thompson, 'The Business of Agriculture: Prairie Farmers and the Adoption of "Business Methods," 1880–1950,' in *Canadian Papers in Business History,* vol. I (1989), 245–69, and the works on Ontario and Quebec referred to in notes 26 and 27 above.

The revelation that only one farm in six in Canada West produced significant surpluses in 1861 must be considered. The participation of even remote bush farmers in local markets – a few eggs and some potash for sale, tea and molasses and axes for purchase – demonstrates that the other five (of six) were more than subsistence operations. For most of these households, as for Elizabeth Goudie, wage labour was very important in certain seasons, in certain phases of a farm's development (such as timber clearing), and as an income supplement. However, the use of markets and participation in wage labour do not stand out as the dominant characteristics of these farms.

The most valuable input to farm production came from the family itself. Thus the pioneer farm participated in the emerging global capitalist system on a part-time basis at best. Using the not-quite-analogous term of 'peasant agriculture,' Eric Hobsbawm suggests that non-capitalist farm families endured long after the founding of factories and long after most historians had adopted the concepts and categories associated with the Industrial Revolution in their discussions of North Atlantic society. Hobsbawm argues that the peasantry ceased to be a class in rural France only in the twentieth century: 'Between the time when the bulk of the peasantry, i.e. the bulk of the pre-industrial population, exists largely outside the capitalist market, and the time when the bulk of the surviving agricultural population can be regarded as "capitalist," there is a long period when we can claim no more than that the peasantry adapts itself to the dominance of a capitalist economy' (Hobsbawm, 'Scottish Reformers,' 22). In Canada, some household economies endured as long as those in France and presented similar conceptual problems for students of social organization.

30 Margaret Conrad, Alvin Finkel, and Cornelius Jaenen, *History of the Canadian Peoples: Beginnings to 1867* (Toronto: Copp Clark Pitman 1993), 502.

Chapter 4: Family Chains and Thunder Gusts

1 Elizabeth Goudie, *Woman of Labrador*, ed. David Zimmerly (Halifax: Nimbus Publishing 1996).

2 In one chapter, Elizabeth wrote: 'After the ... oice complete froze over I went to catch trout for my winter I walked 4 miles and carried my food for a week and i catched trout I catched about 5 Hundred.' In the summer they headed for the salmon fishing grounds in their fourteen-foot boat, a trip of ninety miles, and there found many other families and plenty of fish that could be dried and stored for the winter's needs. At the end of the fishing season, 'we all went to Rigolet for the big round up for the year the custom of the fishermen when the fishing season was over they always had a big dance in rigolet its the oldest H[udson's] B. Bay post on Labrador so we spend t[w]o or three nights dancing ... [Then] every one broke up for another year ... went to th[eir] winter homes and started pre[par]ing for winter.' This version of the notebooks differs in spelling and language from Zimmerly's edition. It was prepared

by Gerald M. Sider, *Culture and Class in Anthropology and History* (Cambridge: Cambridge University Press 1986), 158–61.

3 Gerald L. Pocius, *A Place to Belong: Community Order and Everyday Space in Calvert, Newfoundland* (Montreal: McGill-Queen's University Press 1991), 42–3.

4 A survey of this theme is Walter J. Ong, *Orality and Literacy: The Technologizing of the Word* (London: Methuen 1982).

5 Rosalind Thomas, *Literacy and Orality in Ancient Greece* (Cambridge: Cambridge University Press 1992), 129, 89. Brian Stock describes 'textual communities' of non-readers who, because of their knowledge of given texts, are actually 'literate' after a fashion. He suggests that this 'lay literacy' can be defined as one in which people know that texts exist, can be consulted, have an invariable meaning, must be interpreted correctly, and thus have an authority that transcends the actual reader or speaker; Brian Stock, *The Implications of Literacy: Written Language and Models of Interpretation in the Eleventh and Twelfth Centuries* (Princeton, NJ: Princeton University Press 1983), 88–91.

6 This was the baby who died shortly after birth.

7 The reference to 'the Government' occurs in the 1930s, a decade known in a number of countries for the necessity of government relief payments. Sean Cadigan, 'Battle Harbour in Transition: Merchants, Fishermen, and the State in the Struggle for Relief in a Labrador Community during the 1930s,' *Labour/Le Travail* 26 (fall 1990), 125–50; Jim Overton, 'Public Relief and Social Unrest in Newfoundland in the 1930s: An Evaluation of the Ideas of Piven and Cloward,' in Gregory S. Kealey, ed., *Class, Gender, and Region: Essays in Canadian Historical Sociology* (St John's 1988), 153–66.

8 J. Goody and I. Watt, 'The Consequences of Literacy,' in J. Goody, ed., *Literacy in Traditional Societies* (Cambridge: Cambridge University Press 1968), 58.

9 David Vincent, *Literacy and Popular Culture: England 1750–1914* (Cambridge: Cambridge University Press 1989), 228–69.

10 Terence Crowley, '"Thunder Gusts": Popular Disturbances in Early French Canada,' Canadian Historical Association *Historical Papers* (1979), 11–32. To say that such an approach deals with the politics of household economies or family-based cultures is to employ the term 'politics' in a special way. It implies that the central themes of political studies in these distinctive communities will be power and hegemony, resistance and subordination. The essence of politics for

these groups of ordinary people lies in the degree of domination that they must accept. Though far from identical, elements of their political life resemble those of such groups as slaves, serfs, and lower castes.

James C. Scott has suggested that each of these categories 'represents an institutionalized arrangement for appropriating labor, goods, and services from a subordinate population. As a formal matter, subordinate groups in these forms of domination have no political or civil rights, and their status is fixed by birth. Social mobility, in principle if not in practice, is precluded. The ideologies justifying domination of this kind include formal assumptions about inferiority and superiority which, in turn, find expression in certain rituals or etiquette regulating public contact between strata. Despite a degree of institutionalization, relations between master and slave, the landlord and the serf ... are forms of personal rule providing great latitude for arbitrary and capricious behavior by the superior ... Finally, subordinates in such large-scale structures of domination nevertheless have a fairly extensive social existence outside the immediate control of the dominant.'

Scott addresses issues of dignity and autonomy, which can be examined not on the hustings or in the court but in informal settings. Thus his material lies in 'the rumors, gossip, folktales, songs, gestures, jokes and theater of the powerless as vehicles by which, among other things, they insinuate a critique of power while hiding behind anonymity or behind innocuous understandings of their conduct.' The same type of source will reveal comparable insights into the households of early northern North America because these politics too were the politics of informal resistance; James C. Scott, *Domination and the Arts of Resistance: Hidden Transcripts* (New Haven, Conn.: Yale University Press 1990), xi, xiii.

A useful introduction to this literature in the context of English history is R.W. Malcolmson, *Life and Labour in England 1700–1780* (London: Hutchinson 1981); on the Canadian equivalents, the opening essays in Victor L. Russell, ed., *Forging a Consensus: Historical Essays on Toronto* (Toronto: University of Toronto Press 1984), by Romney, Kealey, Dyster, and Rogers suggest the dimensions of a people's movement and hint at the dimensions of the popular society that was being left behind. Bryan Palmer's unpublished 'Paternalism, Patriarchy, Power and Class: The Theatre of Politics in Upper Canada in the 1830s' (History Department, Queen's University, Kingston, Ont.) and Colin Read, *Rising in Western Upper Canada 1837–8: The Duncombe Revolt and After* (Toronto: University of Toronto Press 1982), suggest the possibilities of this approach, as does Fred Landon, 'The Common Man in the Era of the Rebellion in Upper Canada,' Canadian Historical Association *Annual Report* (1937), 76–91; a recent work is Allan Greer, *The Patriots and the People: The Rebellion of 1837 in Rural Lower Canada* (Toronto: University of Toronto Press 1993).

11 I propose these interpretations in two essays in my *River Road: Essays on Manitoba and Prairie History* (Winnipeg: University of Manitoba Press 1996), 17–22 and 79–90. The intense experience of the uprisings, the influential figure of Riel, and the roots in another cultural world begin to account for the continuity of Métis identity in Canada. Certainly, as section 35 of the Constitution Act, 1982, makes clear, the Métis have survived. They find their active loyalists among the descendants of hunters and traders and freighters who never found prosperity. Their selective tradition celebrates a lost age of freedom and abundance, laments the crisis of buffalo extinction and the hanging of Riel, and struggles against the community's impoverishment in the twentieth century. This collective consciousness is powerful. The Métis themselves laboured diligently to interview elders and to ensure the publication of a book, A.H. de Trémaudan, *Histoire de la Nation Métisse dans l'Ouest Canadien* (1936), in which they might preserve their interpretation of the past. They thus maintained a political identity outside the established order in Canada, grounded in an economy and in cultural expressions that grew out of the land.

12 Sider, *Culture and Class*, 88.

13 On the Newfoundland economy, see Sean T. Cadigan, 'Merchant Capital, the State, and Labour in a British Colony: Servant–Master Relations and Capital Accumulation in Newfoundland's Northeast-Coast Fishery, 1775–1799,' *Journal of the Canadian Historical Association* 2 (1991), 17–42; John E. Crowley, 'Empire versus Truck: The Official Interpretation of Debt and Labour in the Eighteenth-Century Newfoundland Fishery,' *Canadian Historical Review* 70, no. 3 (1989), 311–36; Rosemary E. Ommer, ed., *Merchant Credit and Labour Strategies in Historical Perspective* (Fredericton: Acadiensis Press 1990); Ommer, *From Outpost to Outport: A Structural Analysis of the Jersey–Gaspé Cod Fishery 1767–1886* (Montreal: McGill-Queen's University Press 1991); and Ommer, 'Merchant Credit and the Informal

Economy: Newfoundland, 1918–1929,' Canadian Historical Association *Historical Papers* (1989), 167–89. A survey is Peter Neary and Patrick O'Flaherty, *Part of the Main: An Illustrated History of Newfoundland and Labrador* (St John's: Breakwater 1983); the modern political context is presented in Peter Neary, *Newfoundland in the North Atlantic World, 1929–1949* (Montreal: McGill-Queen's University Press 1988); also Peter Sinclair, *From Traps to Draggers: Domestic Commodity Production in Northwest Newfoundland, 1850–1982* (St John's: Institute of Social and Economic Research, Memorial University of Newfoundland, 1985).

14 Linda Little, 'Collective Action in Outport Newfoundland: A Case Study from the 1830s,' *Labour/Le Travail* 26 (fall 1990), 7–35; Cadigan, 'Battle Harbour in Transition'; Ian McKay, 'Historians, Anthropology, and the Concept of Culture,' *Labour/Le Travail* 8/9 (1981–2), 205–11; Ian McDonald, *'To Each His Own': William Coaker and the Fishermen's Protective Union in Newfoundland Politics, 1908–1925*, ed. J.K Hiller (St John's: Institute of Social and Economic Research, Memorial University of Newfoundland, 1987).

15 The great Maritime evangelist Henry Alline outlined these truths in a hymn concerning his own miraculous conversion to the faith:

No mortal tongue can ever tell,
The horrors of that gloomy night,
When I hung o`er that brink of hell,
Expecting soon my wretched flight!

I felt my burden waste my life,
While guilt did ev`ry hope devour,
Trembling I stretch`d with groans and strife
For to escape the dreadful hour.

But in the midst of all my grief,
The great Messiah spoke in love;
His arm appeared for my relief,
And bid my guilt and sorrow move.

He pluck`d me from the jaws of hell,
With his almighty arm of pow`r;
And O! no mortal tongue can tell,
The change of that immortal hour!

Then I enjoy'd a sweet release,
From chains of sin and pow'rs of death,
My soul was fill'd with heav'nly peace,
My groans were turn'd to praising breath.

From H. Alline, *Hymns and Spiritual Songs* (Boston 1786), quoted in
G.A. Rawlyk, *Champions of the Truth: Fundamentalism, Modernism,
and the Maritime Baptists* (Montreal: McGill-Queen's University
Press 1990), 9.
16 William Westfall, *Two Worlds: The Protestant Culture of Nineteenth-
Century Ontario* (Montreal: McGill-Queen's University Press 1989),
38–43.
17 Rawlyk, *Champions of the Truth*, 21.
18 Ibid., 33.
19 Henry Alline's report on a meeting at Liverpool, Nova Scotia, in
1782 offers an example: 'Almost all the town assembled together,
and some that were lively christians prayed and exhorted, and
God was there with a truth. I preached every day, and sometimes
twice a day; and the houses where I went were crowded almost
all the time. Many were brought out of darkness and rejoiced, and
exhorted in public. And O how affecting it was to see some young
people not only exhort their companions, but also take their parents by the hand, and entreat them for their soul's sake to rest no
longer in their sins, but fly to Jesus Christ while there was hope.
One young lad ... I saw, after sermon, take his father by the hand,
and cry out, "O father, you have been a great sinner, and now are
an old man: an old sinner, with grey hairs upon your head, going
right down to destruction. O turn, turn, dear father, return and fly
to Jesus Christ"' (my emphasis); Rawlyk, *Champions of the Truth*,
12, quoting J. Beverly and B. Moody, eds., *The Life and Journal of
the Rev. Mr. Henry Alline* (Hantsport, NS: Lancelot Press 1982),
208–9.
     Donald G. Mathews has suggested that 'the revolutionary qual-
ity' of the evangelical movement 'was not its assault upon power,
for it made none, but its weakening of the cultural, religious, and
psychological constraints upon people of relatively low estate by
elevating them in their own esteem and giving them the personal
discipline to use their lives as best they could in Christian service.'
Cited in Michael Gauvreau, 'Protestantism Transformed: Personal
Piety and the Evangelical Social Vision, 1815–1867,' in G.A. Raw-

lyk, ed., *The Canadian Protestant Experience 1760 to 1990* (Burlington: Welch 1990), 49.

20 I would like to thank Geoff Sherington and Chad Gaffield for their assistance on this subject; the issue is surveyed effectively in Chad Gaffield, 'Children, Schooling, and Family Reproduction in Nineteenth-Century Ontario,' *Canadian Historical Review* 72, no. 2 (1991), 157–91; also Allan Greer and Ian Radforth, eds., *Colonial Leviathan: State Formation in Nineteenth-Century Canada* (Toronto: University of Toronto Press 1992); Bruce Curtis, *Building the Educational State: Canada West 1836–1871* (London: Althouse Press 1988), and his *True Government by Choice Men? Inspection, Education, and State Formation in Canada West* (Toronto: University of Toronto Press 1992); Tamara Hareven and Andrejs Plakans, eds., *Family History at the Crossroads: A Journal of Family History Reader* (Princeton, NJ: Princeton University Press 1987); and R.M. Netting, et al., *Households: Comparative and Historical Studies of the Domestic Group* (Berkeley: University of California Press 1984). This topic is discussed at greater length below in chapter 6.

Chapter 5: Phyllis Knight and Canada's First Century

1 Phyllis Knight and Rolf Knight, *A Very Ordinary Life* (Vancouver: New Star Books 1974). I would like to thank Rolf Knight for his permission to quote from this remarkable memoir, for several valuable conversations about his family, and for his careful reading of these chapters and his suggestions about my interpretation of his mother's experience.

2 John Porter, *The Vertical Mosaic: An Analysis of Social Class and Power in Canada* (Toronto: University of Toronto Press 1965), 29–59.

3 Robert Heilbroner, *The Nature and Logic of Capitalism* (New York: Norton 1987), 134.

4 Wealth found its measure, too, in a commonly-accepted *language* of money. Thomas Carlyle, the English writer, commented on the increasingly-powerful 'cash nexus' which, by the 1830s, had become the dominant force in social relationships to the exclusion of older ties of mutual obligation. This argument was resumed by the historian E.P. Thompson in the 1960s and 1970s in his discussions of time and work discipline; his essays are collected in *Customs in Common: Studies in Traditional Popular Culture* (New York:

The New Press 1993). This discussion relies, too, on Robert Heilbroner, *The Nature and Logic of Capitalism*.

5 Roy Vogt, *Whose Property? The Deepening Conflict between Private
Property and Democracy in Canada* (Toronto: University of Toronto
Press 1999).

6 Alan McCullough, *The Commercial Fishery of the Canadian Great Lakes*
(Ottawa: Minister of Supply and Services 1989), 68–9.

7 Another means of understanding the shift in social relations, as the
new social history would depict it, is to focus on profit, which takes
three forms. First, the grocery store, for example, charges more for
the sugar it sells than the price it paid to the wholesaler; this type of
profit accrues in the very act of exchange. In a second type, represented by the owner and wage labourer in a sugar refinery, the difference between labour and other input costs and the price of the
product accrues to the owner. The surplus is generated within the
production system itself and is said (particularly by Marxists) to
represent surplus value drawn from the sweat of the workers themselves. A third form of profit, described by the economist Joseph
Schumpeter, stems from innovation such as the development of the
sugar replacement Aspartame. In this case, capitalists enjoy a (theo-
retically short-lived) monopoly when they create a new product or
process that captures larger than 'normal' gains for a brief period
before competitors catch up.

The three forms of profit all rely on popular acceptance of property definitions. They also assume that owners of capital, not
labour, should possess whatever surplus ensues. Although the surpluses originate in trade, or production, or innovation, their crucial
characteristic lies in a collective political decision about property
rights. Such a decision establishes distinct categories of citizens,
based on owning or failing to own capital.

8 Nevertheless, the value of farm production increased by a multiple
of four between 1870 and 1925. Cheese, butter and hogs in central
Canada supplemented prairie wheat and the BC and Maritime
crops of apples, potatoes, and livestock as farm contributors to
national economic growth. See R.M. McInnis, 'Perspectives on
Ontario Agriculture 1815–1930,' *Canadian Papers in Rural History*,
vol. VIII (Gananoque: Langdale Press 1992), 17–127. Agriculture's
share of gross domestic product (GDP) dropped from 40 per cent to
22 per cent between 1870 and 1910. However, the value of farm production in real terms still remained a significant proportion of

Canadian wealth. According to Marvin McInnis, gross value added (GVA) in agriculture rose from about $136 million in 1870–1 to $568 million in 1926–7 (constant 1913 dollars), or about four times in fifty-five years (7.6 per cent annually, on average). Also useful are M.C. Urquhart's analyses in 'Canadian Economic Growth 1870–1980,' Queen's University Department of Economics Discussion Paper No. 734, 1988, for which reference I wish to thank Ken Norrie. Also helpful is Kenneth Norrie and Douglas Owram, *A History of the Canadian Economy* (Toronto: Harcourt Brace Jovanovich 1991), 317.

9 Together the natural resources, including old and new staples, accounted for about 10 per cent of Canada's wealth production in 1900. Norrie and Owram, *Canadian Economy*, 336, offer estimates of this value.

10 Iron/steel, food/beverage, clothing/textiles, and wood products dominated the manufacturing sector (60 per cent of the total), but a large number of smaller industries in Saint John, Montreal, Toronto, Hamilton, and other growing cities also figured in the country's output. The periods of rapid industrial expansion – the 1880s, 1900–12, 1915–18, 1924–9 – consolidated Canada's claim to be a genuine industrial power, in this view.

11 Norrie and Owram, *Canadian Economy*, 358, 553.

12 Urquhart, 'Canadian Economic Growth 1870–1980,' 8–10; also see Urquhart's discussion in his essay 'New Estimates of Gross National Product, Canada, 1870–1926: Some Implications for Canadian Development,' in Stanley L. Engerman and Robert E. Gallman, *Long-Term Factors in American Economic Growth* (Chicago: University of Chicago Press 1986).

13 Economic Council of Canada, *Living Together: A Study of Regional Disparities* (Ottawa: Minister of Supply and Services 1977) and *Western Transition* (Ottawa: Minister of Supply and Services 1984).

14 Three in ten households still lived and worked in farming, fishing, woods-working and trapping activities, or some combination thereof, in 1940. Inevitably, these family members saw the world quite differently than did their settler ancestors. However, as Elizabeth Goudie's memoir shows, a new layer of language and experience merged only slowly with, and did not obliterate, the earlier one.

About four in ten Maritime familes still were fishery- or farm-based producers in 1901 and three in ten retained that role, or com-

bined these activities, in 1941. [David Alexander says that 31 per
cent of the Maritime labour force was engaged in primary production in 1941; Alexander, 'Economic Growth in the Atlantic Region,'
in David Bercuson and Phillip Buckner, eds., *Eastern and Western
Perspectives: Papers from the Joint Atlantic Canada/Western Canadian
Studies Conference* (Toronto: University of Toronto Press 1981), 202;
Ernest Forbes says that 62 per cent lived in rural areas in 1931:
Forbes, ed., *Challenging the Regional Stereotype: Essays on the Twentieth Century Maritimes* (Fredericton: Acadiensis Press 1989), 151.

About four in ten Quebec and Ontario families were engaged in
primary production (agriculture, fishing, and trapping) in 1901, and
two in ten had this status in 1941; Alexander, 'Economic Growth,'
202, Table 1. Nearly half of western Canadians were farm-dwellers
throughout this era; Gerald Friesen, *The Canadian Prairies: A History,*
student edition (Toronto: University of Toronto Press 1987), Tables 7
and 8, 514–17.

15 In North America, the fluid market for labour in the wider community – a capitalist labour market – ensured maintenance of family
farms by supplying or absorbing extra workers, depending on the
season and the ages and gender of family members. Because of the
kinship basis of labour relations – that is, individuals were moti-
vated to contribute their labour to the household out of a sense of
obligation to the family itself – and the absence of 'exploitation' of
one class by another in the formal Marxist sense of the word, Friedmann insists that farm households such as those in prairie Canada
are not properly described as capitalist. A similar case could be
made for Atlantic and Quebec households, where a variety of economic activities sustained the family. This approach is discussed in
Harriet Friedmann, 'World Market, State, and Family Farm: Social
Bases of Household Production in the Era of Wage Labor,' *Comparative Studies in Society and History* 20, no. 4 (1978), 545–86; also Friedmann, 'Simple Commodity Production and Wage Labour in the
American Plains,' *Journal of Peasant Studies* 6, no. 1 (1978), 71–100,
and 'Household Production and the National Economy: Concepts
for the Analysis of Agrarian Formations,' *Journal of Peasant Studies*
7, no. 2 (1980), 158–84; it is also presented in Jeffery Taylor, *Fashioning Farmers: Ideology, Agricultural Knowledge, and the Manitoba Farm
Movement, 1890–1925* (Regina: Canadian Plains Research Centre
1994).

16 Canadian prairie agriculture took shape between 1880 and 1914, as

sweeping changes transformed the global economy. Extraordinary as it seems in retrospect, the western homestead was part of a 'new world-wide division of economic functions.' The allegedly empty lands – if one ignored Aboriginal peoples, as the Europeans did – in Australasia, eastern Europe, and the Americas were converted into agricultural zones where goods were produced for export. On the Canadian prairies, transplanted Europeans and Americans, as well as Canadians, all of whom were accustomed to the rule of price and market and private property, set up homes and cultivated their fields in order to meet Europe's food needs. These immigrant homesteaders sold grain on a truly international market. For this simple reason, the farm households *seemed* to be capitalist: North American frontier rhetoric made much of their independence and profit-seeking while celebrating their ownership of the means of production. Is it any wonder that farmers and their families assumed that they would have little in common with wage-earners in factory towns? The quotation is taken from S.B. Saul, *Studies in British Overseas Trade 1870–1914*, cited in Donald Denoon, *Settler Capitalism: The Dynamics of Dependent Development in the Southern Hemisphere* (Oxford: Clarendon Press 1983), 212.

17 An introduction to the two approaches might be to compare Paul Voisey, *Vulcan: The Making of a Prairie Community* (Toronto: University of Toronto Press 1988), with Taylor, *Fashioning Farmers*.

Some recent scholarly writing would paint the farm as capitalist, but only in the case of one member of the household – the male 'head' of the family. The very idea of a family economy suggests that all members contributed to a single enterprise. None the less, in almost every Canadian farm community, the male head of the family alone legally owned the enterprise. As Marjorie Griffin Cohen has argued, 'Ownership and control [are] as significant in the family economy ... as in more identifiably capitalist relationships.' Cohen demonstrates that men regularly chose to invest the farm's resources in their own work, leaving women's tasks to the primitive labour systems of yesteryear. And in her Ontario investigation she discovered that when the market for the women's products, especially milk, cheese and butter, increased substantially, men took control of the entire operation and rapidly mechanized this 'profit centre.' Is it any wonder that some of the best-known farm novels – M. Ostenso's *Wild Geese* (Toronto: McClelland and Stewart 1925), F.P. Grove's *Settlers of the Marsh* (Toronto: Ryerson Press 1925; Tor-

onto: McClelland and Stewart 1965), and *Fruits of the Earth* first pub.
1933 (Toronto: McClelland and Stewart 1965), and R. Stead's *Grain*
(Toronto: McClelland and Stewart 1926; 1963) drew their power
from concerns about male priorities? Marjorie Griffin Cohen,
*Women's Work, Markets and Economic Development in Nineteenth-Cen-
tury Ontario* (Toronto: University of Toronto Press 1988), 43.

Cohen concluded that women and children were the proletariat
of the family farm. By extension her interpretation makes the male
farmer a capitalist or bourgeois exploiter. This argument compli-
cates the historical problem of how to generalize about family relationships and underestimates the degree to which all family
members accepted and lived within a capitalist environment. It is
true, however, that the farm's property structure translated the
inevitable disputes arising in daily work into the quarrels of a marriage relationship.

Recent research on prairie farm families takes a different tack.
Sarah Brooks Sundberg argues that the image of prairie farm wife
as 'helpmate' underestimates the contributions and power of such
women. Royden Loewen demonstrates that, in the case of one
group of prairie Mennonites, property was distributed equally
between male and female heirs. In her discussion of prairie
women's 'double load' of labour, Veronica Strong-Boag emphasizes
that women waged a remarkable struggle for changes in their daily
lives *and* in the public estimation of their work. Simultaneously
they accepted the endless domestic responsibilities that accompanied the job of homemaker. Their struggle – in the home workplace
and in public forums – itself bore witness to their ability to shape
their own circumstances.

In the absence of further evidence, it does not seem fruitful to
claim that family is analogous to factory, that father (in the family)
and boss (in wage-labour settings) are synonymous, that household
discipline is the same as wage discipline, and that the male farmer
is by definition a member of Karl Marx's bourgeoisie. All members
of the farm household, as farmers and as family, though undeniably
affected by assumptions about the social implications of sexual difference, were shaped in similar degree by the increasingly capitalist
context in which they lived. Cohen, *Women's Work*, 44, 116; Sarah
Brooks Sundberg, 'Farm Women on the Canadian Prairie Frontier:
The Helpmate Image,' in Veronica Strong-Boag and Anita Clair
Fellman, eds., *Rethinking Canada: The Promise of Women's History*

(Toronto: Copp Clark Pitman 1986), and 'A Female Frontier: Manitoba Farm Women in 1922,' *Prairie Forum* 16, no. 2 (1991), 185–204; Royden Loewen, '"The Children, the Cows, My Dear Man and My Sister": The Transplanted Lives of Mennonite Farm Women 1874–1900,' *Canadian Historical Review* 73, no. 3 (1992), 344–73; and Veronica Strong-Boag, 'Pulling in Double Harness or Hauling a Double Load: Women, Work and Feminism on the Canadian Prairie,' *Journal of Canadian Studies* 21, no. 3 (1986), 32–52.

18 Two approaches to these issues are Joan Wallach Scott, *Gender and the Politics of History* (New York: Columbia University Press 1988), and Joy Parr, *The Gender of Breadwinners: Women, Men, and Change in Two Industrial Towns 1880–1950* (Toronto: University of Toronto Press 1990).

19 Terry Copp, *The Anatomy of Poverty: The Condition of the Working Class in Montreal, 1897–1929* (Toronto: McClelland and Stewart 1974); Bryan Palmer, *Working-Class Experience: Rethinking the History of Canadian Labour, 1800–1991*, 2nd ed. (Toronto: McClelland and Stewart 1992), 192; but see Paul Rutherford's disagreement in *A Victorian Authority: The Daily Press in Late Nineteenth-Century Canada* (Toronto: University of Toronto Press 1982), 247 n. 43.

20 Bettina Bradbury, *Working Families: Age, Gender, and Daily Survival in Industrializing Montreal* (Toronto: McClelland and Stewart 1993).

21 Bradbury, *Working Families*; Veronica Strong-Boag, 'Keeping House in God's Country: Canadian Women at Work in the Home,' in Craig Heron and Robert Storey, eds., *On the Job: Confronting the Labour Process in Canada* (Montreal: McGill-Queen's University Press 1986).

22 The approach is best expressed in Donald Creighton, *Canada's First Century 1867–1967* (Toronto: Macmillan 1970).

23 A special issue of *Labour/Le Travail* 32 (1996), which compares Canada and Australia, makes this point repeatedly in its discussion of Canada.

Donald Creighton changed his view of class between 1941 and 1956. In the former year he wrote, in connection with Canadian experience in the First World War: 'Before the war, Canada had believed herself to be a land of equality of opportunity; but the war increased the social stratification of the country and deepened the class consciousness of its component groups.' Creighton, 'Federal Relations in Canada since 1914,' in Chester Martin, ed., *Canada in Peace and War: Eight Studies in National Trends since 1914* (Toronto: Oxford University Press 1941), 42; but when he was president of the

Canadian Historical Association he rejected class analysis; Creighton, 'Presidential Address,' Canadian Historical Association *Historical Papers 1957*, 6–8.

24 Palmer, *Working Class Experience*, 214–17; a different perspective is
presented in Michael Bliss, *Northern Enterprise: Five Centuries of
Canadian Business* (Toronto: McClelland and Stewart 1987); the story
is the focus of Wallace Clement, *Continental Corporate Power: Eco-
nomic Elite Linkages between Canada and the United States* (Toronto:
McClelland and Stewart 1977); also Trevor J.O. Dick, 'Canadian
Newsprint, 1913–1930: National Policies and the North American
Economy,' in Douglas McCalla, ed., *Perspectives on Canadian Eco-
nomic History* (Toronto: Copp Clark Pitman 1987); H.V. Nelles, *The
Politics of Development: Forests, Mines and Hydro-Electric Power in
Ontario, 1849–1941* (Toronto: Macmillan 1974); and Tom Traves, *The
State and Enterprise: Canadian Manufacturers and the Federal Govern-
ment, 1917–1931* (Toronto: University of Toronto Press 1979).

25 Norrie and Owram, *Canadian Economy*, 335–57; Kari Levitt, *Silent
Surrender: The Multinational Corporation in Canada* (Toronto: Mac-
millan 1970); Larry Pratt, 'The State and Province-Building:
Alberta's Development Strategy,' in Leo Panitch, ed., *The Canadian
State: Political Economy and Political Power* (Toronto: University of
Toronto Press 1977), 133–62; Nelles, *The Politics of Development*.

26 Craig Heron, 'The New Factory Regime and Workers' Struggles in
Canada, 1890–1940,' and Neil Sutherland, '"We always had things
to do": Anglophone Children in Urban Household and Family
Economies between the 1920s and the 1960s,' unpublished papers
presented to the Australia–Canada Labour Studies Conference,
University of Sydney, 1989; and Heron, 'The High School and the
Family Economy in a Factory City: Hamilton, 1890–1940,' paper
presented to Canadian Historical Association, Victoria, 1990.

27 E.P. Thompson, 'Time, Work-Discipline and Industrial Capitalism,'
in Thompson, *Customs in Common: Studies in Traditional Popular Culture* (New York: New Press 1993), 390.

28 All these characteristics, admittedly drawn from American and
British historical literature, marked Canadian experience as well.
Mary Ryan, *Cradle of the Middle Class: The Family in Oneida County,
New York, 1790–1865* (Cambridge: Cambridge University Press
1981); Stuart M. Blumin, 'The Hypothesis of Middle-Class Formation in Nineteenth-Century America: A Critique and Some Propos-
als,' *American Historical Review* (1985), 299–338; Harold Perkin, *The*

Rise of Professional Society: England since 1880 (London: Routledge 1989); Erik Olin Wright, *Classes* (London: Verso 1985); also the essays on class in Eric Hobsbawm, *Worlds of Labour: Further Studies in the History of Labour* (London: Weidenfeld and Nicolson 1984), especially 176–213. For a more recent and poststructural view, see Patrick Joyce, *Class* (Oxford: Oxford University Press 1995).

29 J.M.S. Careless, *The Union of the Canadas: The Growth of Canadian Institutions 1841–1857* (Toronto: McClelland and Stewart 1967), 144–5.

30 These two paragraphs rely on and quote from James W. Carey, 'Technology and Ideology: The Case of the Telegraph,' in Carey, ed., *Communication as Culture: Essays on Media and Society* (New York: Routledge 1989), 220–2.

Chapter 6: Literate Communication and Political Resistance

1 These issues are discussed in Stephen Kern, *The Culture of Time and Space 1880–1918* (Cambridge: Harvard University Press 1983).

2 The notion of two 'founding peoples' is constructed on the print-power of these two tongues, as well as military, economic, and administrative power. The concept ignores the Aboriginal role in Canada's foundation; Benedict Anderson, *Imagined Communities: Reflections on the Origin and Spread of Nationalism* (London: Verso 1991), 15.

3 In her reflections on the completion of the manuscript, Phyllis Knight returned to this theme: 'Those open spaces and the free and easy way of things, the fact that there were pretty few regulations, were the most appealing things about Canada. I know, it was pretty wasteful and it also meant that you had little protection and security. But there was something worthwhile in it. I'm the last person in the world to say that those times were great. But tough times and everything, I'm glad I had a chance to be part of it.' Phyllis Knight and Rolf Knight, *A Very Ordinary Life* (Vancouver: New Star Books 1974), 298.

4 Phyllis mentioned social snubs in the street, the suspicion that neighbours might be spying on them, and several acts of vandalism: 'A few of the people I knew in the neighbourhood weren't affected by that hysteria and we remained friends. They treated you fairly and decided whether they liked you or not as an individual. But there were very few people like that. And they were often peo-

ple that you wouldn't have expected it of either. There was a couple who worked in the travelling carnival and another who was a two hundred percent Englishman. But the war really brought out pettiness and viciousness in what were otherwise decent ordinary people.' Ibid., 224.

5 Rolf Knight to author, 12 June 1998.

6 Anderson, *Imagined Communities*, 35–6.

7 Anderson, *Imagined Communities*, and Fernand Ouellet, 'Le Nationalisme canadien-francais: De ses origines à l'insurrection de 1837,' *Canadian Historical Review* 45, no. 4 (1964), 277–92. Print-capitalism – the marriage of printing technology and capitalist economic forces – first touched northern North America in the mid-eighteenth century, and by the early nineteenth century printer-journalists fuelled many public debates. Their publications commenced as organs of the bureaucracy and the market-place, providing news of decrees and appointments, mingled with reports on prices and ship arrivals, but their effect was to imagine a new shape for the land. The journalists publicized different perspectives on world and local events than had been available in the metropolitan press – local refractions originating in a 'British North American' site and aimed at the interests of local audiences.

8 Harvey J. Graff, 'Literacy and Social Structure in Elgin County, Canada West: 1861,' *Histoire Sociale/Social History* 6, no. 1 (April 1973), 25–48; Allan Greer, 'The Pattern of Literacy in Quebec, 1745–1899,' *Histoire sociale/Social History* 11, no. 2 (1978), 293–335; Jean de Bonville, *La Presse québécoise de 1884 à 1914: Genèse d'un média de masse* (Quebec: Laval University Press 1988), 14–16; Susan E. Houston and Alison Prentice, *Schooling and Scholars in Nineteenth-Century Ontario* (Toronto: University of Toronto Press 1988), 15–17, 84–5; David G. Alexander,' 'Literacy and Economic Development in Nineteenth-Century Newfoundland' in his *Atlantic Canada and Confederation: Essays in Canadian Political Economy* (Toronto: University of Toronto Press 1983), 110–43.

On one level, literacy is simply a tool – the ability to read and write. One might imagine that it is an individual accomplishment; one might define it *only* as the ability to read and write; one might assume that personal qualities, including intelligence and perseverance, have ensured that some people read effectively and critically, whereas others read badly or not at all; one might even generalize about the virtues possessed by social *groups* that use print effectively and those that rarely consult the printed word. But intensive study of the impact of literacy on mental processes has not sustained the thesis that it underlies a cognitive revolution distinguishing the thought processes of literate individuals from those of non-literate individuals.

Many scholars of literacy are now less inclined to make broad assumptions about 'modernized' minds. They argue instead that literacy and orality are ways of thinking. They define literacy not simply as reading or writing but as a kind of thinking in which one distinguishes between *language* (the tongue itself – the structures of Cree and French, for example – as well as oral and written modes of expression), *text* (the assumed content or message, whether in a speech or a book or in the land around us), and *interpretation* (including the perspectives of the sender and of the receiver). They suggest that people employ both literate and oral approaches to experience, and infinite combinations thereof, in their daily lives. Literacy, as opposed to orality, demands that people objectify what they are thinking about and thus enables them to make 'self-conscious distinctions ... between language structure, discourse meanings, and interpretations.' Judith A. Langer, 'A Sociocognitive Perspective on Literacy,' in Langer, ed., *Language, Literacy, and Culture: Issues of Society and Schooling* (Norwood, NJ: Ablex Publishing 1987), 3; M.M. Bakhtin, *The Formal Method in Literary Scholarship: A Critical Introduction to Sociological Poetics* (Baltimore: Johns Hopkins University Press 1978).

Scholars such as Judith Langer reject Walter Ong's influential thesis that oral thinking is 'emotional, contextualized and ambiguous' whereas literate thinking is 'abstract, decontextualized and logical.' (Langer, *Language, Literacy, and Culture*, 6; she is responding to Walter Ong's fascinating survey, *Orality and Literacy: The Technologizing of the Word* (London: Routledge 1982).

On the relations between work and literacy, see Harvey J. Graff, 'Respected and Profitable Labour: Literacy, Jobs and the Working Class in the Nineteenth Century,' in Gregory S. Kealey and Peter Warrian, eds., *Essays in Canadian Working Class History* (Toronto: McClelland and Stewart 1976), 58–82.

According to David Olson, there were four prerequisites for the emergence of the modern literate society: a device for 'fixing' and accumulating texts (paper and eventually print), institutions that rely on texts (religion, law, bureaucracy, economy, and science),

institutions that could induct learners into these society-wide activities (family, church, and, most of all, school), and a meta-language
for conceptualizing the discussion of the accumulated texts, for
facilitating 'the spiralling change that comes about when people use
their literacy skills to think, rethink, and reformulate their knowledge and their worlds.' (The phrase comes from Langer, *Language,
Literacy and Culture*, 2–3). Olson also argues that literacy and orality
are simply two ways of dealing with the world around us. He
claims that each is appropriate to certain forms of experience and
that each offers valued forms of rationality; David R. Olson, 'Literate Thought,' in Che Kan Leong and Bikkar S. Randhawa, eds.,
*Understanding Literacy and Cognition: Theory, Research, and Applica-
tion* (New York: Plenum Press 1989), 9–13.

Facility with written material, like language and thinking in general, has a social origin. Thus all citizens as individuals are able to
think in the mode of oral discourse, and all (whether or not they can
read or write) are able to reason in the fashion defined as literate
thought. This is not to deny that Canadian society was transformed
during the nineteenth century from one in which oral communication dominated to one in which literacy became paramount. One
can conclude that Canadians acquired literacy and mastered its
accompaniments – 'signs' such as metaphors and history lessons –
to serve social ends. When the need for the new technical tools of
literacy became apparent in their daily lives, people acquired these
tools, thereby re-establishing some control over their environment.
9 Raymond Williams, *Communications* (London: Chatto & Windus
1966), 20.

One approach to these questions is to define them as aspects of
the emerging 'mass media,' a top-down view that focuses on business structure, audience 'reach,' and dominant 'voices.' A different
perspective is to argue that people would have not only to learn to
control the new means of communications but also to further their
understanding of a rapidly changing land. The very existence of the
second line of thought reveals the bias of the first, with its metaphor
'mass communications.' I believe in the virtues of popular control
and popular intellectual growth. Such a perspective recognizes that
people responded to the changing technologies of communication
and expanded their intellectual horizons to encompass the new circumstances of print-capitalism. I am not arguing that it is wrong to
study the modern media as phenomena of mass communication,

but I am saying that this is a one-sided and top-down perspective that must be complemented by a view from the other side – the active minds of ordinary people.

The newspaper played a crucial role in Canadian public communication during this era. It was one force in the shaping of a people's consciousness, and it created untold numbers of Canadian myths. The image of the press as a 'fourth estate,' serving the public interest by conducting free debate and acting as the 'marketplace of ideas' might well have described the reality of mid-nineteenth-century Canada. When presses were small and circulation was numbered in the hundreds or a few thousand, newspapers could be made catalysts for radical change by one or two energetic individuals.

However, Canadian journalism experienced dramatic alterations between the 1870s and the 1930s; to depict the later, larger, daily journals as the imaginings of one mind or to imply that they served as a market-place for public debate, to argue that their news and editorial columns constituted a no-holds-barred ideological struggle, was simply not true. By 1900, or at the latest 1920, the great daily newspapers in any city numbered only a handful. Their balance sheets reflected huge investments that could not be risked for the sake of a quixotic crusade. The journalistic traditions of independence and individuality now were constrained by advertisers, party donors, and a new professional standard of journalism that eventually enshrined an ideal of social responsibility. Williams, *Communications*, 19–20; perhaps William's view is based on Henri Lefebvre, *Critique of Everyday Life*, vol. I (Paris: L'Arche 1947; 1958; London: Verso 1991), 138–75.

Of this era in journalism, Rolf Knight said that workers' 'views of commercial newspapers as conveyors of information about the world were balanced with a belief that they were institutionalized liars ... It was not newspapers or national journals which were Phyllis' – or their – guide to the broader world, but rather *certain kinds of books* – which were more wide-ranging and uncensored than the daily press.' Rolf Knight to author, 12 June 1998.

The people's communication networks represent a challenging subject of study. The communication scholar Harold Lasswell anticipated the manifold problems when he explained that the questions to ask about the media included: 'Who? Says what? To whom? In which channel? With what effect?' In sketching the location of the people's vehicles for communication or conversation one can sug-

gest that distinct popular responses to an overarching capitalist, literacy-based society were continuously being shaped and reshaped;
this discussion begins with Harold Lasswell, 'The Structure and
Function of Communications in Society,' in Lyman Bryson, ed., *The
Communication of Ideas* (New York: Cooper Square c. 1948), 1–24;
cited in Paul Rutherford, *A Victorian Authority: The Daily Press in
Late Nineteenth-Century Canada* (Toronto: University of Toronto
Press 1982), 6; a recent survey is John Fiske, *Introduction to Communication Studies*, 2nd ed., (London: Routledge 1990).

It is tempting to suggest that issues of communication are more
difficult today, thereby implying that no previous society presents
the complexities of cultural analysis that confront us in the age of
literacy. But who among us has experienced the complexities associated with divining animal movements in Aboriginal societies or
of conjuring with the spirits that inhabited European magic?

A helpful introduction is Raymond Williams, 'The Press and Popular Culture: An Historical Perspective,' in George Boyce, James
Curran, and Pauline Wingate, eds., *Newspaper History from the Seventeenth Century to the Present Day* (London: Constable 1978), 41–50.
Also, see Paul Rutherford, *The Making of the Canadian Media* (Toronto: McGraw-Hill Ryerson 1978), 1–44.

Why did the newspaper become so powerful? Paul Rutherford, a
Canadian historian, adapted one of Raymond Williams's ideas in
arguing that the influential Victorian press in Canada and its successors during the next four decades flourished because the wider
society needed such a cultural instrument: 'The emerging bourgeois
democracy required an increasing level of mass involvement in the
daily round of life. People at the top, notably in politics and business, had to be able to communicate with the publics below. Like-
wise, these publics needed some agency to voice their concerns and
so influence the establishment. But, most important, the emerging
society demanded a citizenry with sufficient knowledge to make
reasonable decisions about how to vote, what to buy, where to
invest, how to find jobs, even what to believe. Older institutions
such as the church or the school could not adequately serve all
these ends. The popular press could. It commanded the power to
transmit facts, ideas, and fantasy to so much of the citizenry.' Rutherford, *A Victorian Authority*, 232.

Advertising provided over 60 per cent of some big dailies' revenue in 1914. Jean de Bonville, a student of Quebec journalism,

emphasizes the primary role of economic forces, notably advertising, in shaping the format and content of the modern urban daily. Although there is some merit in his argument, it is only part of the story. After all, by devoting half of its pages to news, the urban daily acknowledged that it relied on both the citizens' desire for information and their wish to consume. Jean de Bonville, *La presse québécoise*, 326, 347, 361–3; Rutherford, *A Victorian Authority*, 147–9, 156–89. Rutherford suggests that monopoly conditions prevailed by about 1900 (230).

10 Philip Schlesinger, 'The Sociology of Knowledge,' in Herbert Gans, ed., *Deciding What's News*, cited in Rutherford, *A Victorian Authority*, 133; see also 133–8 and 147–89.

11 Rutherford, *A Victorian Authority*, 35, and *The Making of the Canadian Media*, 38.

12 Graeme Patterson offers this definition of myth in *History and Communications: Harold Innis, Marshall McLuhan, the Interpretation of History* (Toronto: University of Toronto Press 1990), 166.

13 Angela E. Davis, '"Country Homemakers": The Daily Lives of Prairie Women as Seen through the Woman's Page of the *Grain Growers' Guide* 1908–1928,' *Canadian Papers in Rural History*, vol. VIII (Gananoque: Langdale Press 1992), 163–74.

14 It has been said that William Aberhart of Alberta employed radio and study groups to similar effect during the 1930s. See also Ian Radforth and Joan Sangster, '"A Link between Labour and Learning." The Workers Educational Association in Ontario, 1917–1951,' *Labour/Le Travail* 8/9 (1981–2), 41–78; Gerald Friesen, 'Adult Education and Union Education: Aspects of English Canadian Cultural History in the 20th Century,' *Labour/Le Travail* 34 (fall 1994); Peter Sandiford et al., *Adult Education in Canada: A Survey* (Toronto: University of Toronto Press 1935); Rosabel Fast, 'The Impact of Sponsorship: The University of Manitoba's Rural Adult Education Program, 1936–1945,' MEd thesis, University of Alberta, 1991; M.R. Welton, 'To Be and Build a Glorious World: The Educational Thought and Practice of Watson Thomson, 1899–1946,' PhD dissertation, University of British Columbia, 1983; Frank Cassidy and Ronald Faris, eds., *Choosing Our Future: Adult Education and Public Policy in Canada* (Toronto: OISE Press 1987); M.R. Welton, ed., *Knowledge for the People: The Struggle of Adult Learning in English-Speaking Canada 1818–1973* (Toronto: OISE Press 1987); Ronald Faris, *The Passionate Educators: Voluntary Associations and the Struggle for*

Control of Adult Education Broadcasting in Canada 1919–1952 (Toronto: Peter Martin 1975); Michael R. Welton, 'Conflicting Visions, Divergent Strategies: Watson Thomson and the Cold War Politics of Adult Education in Saskatchewan, 1944–46,' *Labour/Le Travail* 18 (1986), 111–38. I am indebted to Professor Reginald Edwards for his assistance with this material.

15 Rolf Knight to author, 12 June 1998.

16 Phyllis added: 'We always voted CCF. Here in BC they were pretty good, Ernie Winch and Angus MacInnis and some of the others.' Knight and Knight, *A Very Ordinary Life*. And she then modified this impression: 'Today there's the NDP. But all of their top brass are either lawyers or social workers or professors. If you listen to them talk there isn't a working stiff in the whole lot. I think for many of them "Socialism" is a dirty word. Still, who else is there?' Later, when the NDP government was being castigated in the press, she added: 'When I hear that sort of malarkey and hysteria being drummed up by those robbers who ruled the roost here for so long, then I can guess Barrett [the NDP premier] is doing something right. Their newspapers are up to their old tricks again. If I'm still around next time, I'm going to vote for the NDP no matter what.'

17 Rutherford, *The Making of the Canadian Media*, 35.

18 John English, *The Decline of Politics: The Conservatives and the Party System 1901–1920* (Toronto: University of Toronto Press 1977); the Liberals underwent a similar transition under Laurier and King.

19 Reg Whitaker, *The Government Party: Organizing and Financing the Liberal Party of Canada, 1930–1958* (Toronto: University of Toronto Press 1978); Paul Craven, *'An Impartial Umpire': Industrial Relations and the Canadian State 1900–1911* (Toronto: University of Toronto Press 1980); Alvin Finkel, *Business and Social Reform in the Thirties* (Toronto: Lorimer 1979); English, *Decline of Politics*; Doug Owram, *The Government Generation: Canadian Intellectuals and the State, 1900–1945* (Toronto: University of Toronto Press 1986).

20 Nancy Christie and Michael Gauvreau, *A Full-Orbed Christianity: The Protestant Churches and Social Welfare in Canada 1900–1940* (Montreal: McGill-Queen's University Press 1996).

21 On the theme of self-interest, see the comments in Donald Harman Akenson, *The Irish in Ontario: A Study in Rural History* (Montreal: McGill-Queen's University Press 1984), 82–108, 126–34. He describes the atmosphere as 'entrepreneurial individualism' and judges that, in its 'atomistic model of behaviour' and its absence of

larger allegiance, it was 'non-ideological.' The latter judgment is debatable.

The Catholic and French-language versions of this evolution are different. I would like to thank Louise Déchêne for a conversation on these topics and to note Ronald Rudin, 'Revisionism and the Search for a Normal Society: A Critique of Recent Quebec Historical Writing,' *Canadian Historical Review* 73, no. 1 (1992), 30–61, and *Making History in Twentieth-Century Quebec* (Toronto: University of Toronto Press 1997).

A discussion of the dominant channels of conversation among the common people between 1840 and 1950 might be labelled 'intellectual' or 'social' history, but it would be conducted within the subfields of ideology, cultural studies, theology, and religion, each of which carries significant 'baggage.' I am convinced that both ideology and religion should be brought into the discussion. If one wishes to employ a concept such as ideology, which draws on Marxist literature about production, class, power, and property, one must remember that most nineteenth-century Canadians perceived their world within terms drawn from a quite different language, constructed on Christian foundations. The two approaches, ideology and Christianity, which are not simply opposing visions, must be held in tension if one is to do justice to the outlook of ordinary citizens; Terry Eagleton, *Ideology: An Introduction* (London: Verso 1991).

Recently, religion has been defined as 'systems of beliefs that answer the questions that cultures ask'; William Westfall, *Two Worlds: The Protestant Culture of Nineteenth-Century Ontario* (Montreal: McGill-Queen's University Press 1989), 13. This definition, which sounds like another way of talking about ideology while ignoring God or the supernatural entirely, reflects today's cultural relativism. In the late nineteenth century, when cultural relativism was just beginning to inform studies of human society, J.G. Frazer could define religion as 'a propitiation or conciliation of powers superior to man which are believed to direct and control the course of Nature and of human life.' J.G. Frazer, cited by William P. Alston, 'Religion,' in *Encyclopedia of Philosophy*, 140. Frazer's approach veers to the other extreme, because it posits an extra-terrestrial experience, unknowable by ordinary reckoning, that actually shapes events in this world. Despite its ubiquity among nineteenth-century Canadians, it offers little that today's students can readily grasp. Having juxtaposed the relativist and the supernatural, the ideological and the Christian approaches to belief, how does the historian proceed?

Studies of ideology and religion are very closely intertwined but ultimately address slightly different subjects within the general area of 'culture.' The language of any given community, and of its dominant elites, may use religious concepts or ideological messages in order to sustain the established order in that place. Nevertheless, there is a concern for 'the transcendent' in almost all religions that is not present in most ideologies. The merit of thinking about Canadian people's religion of the era 1830–1900 in the same language that is used to discuss their ideology in the years 1900–50 is that many of the aspirations conveyed in people's expression and speech in the first period reappear in their analysis of the world around them in the latter.

Only in the last century or so have scholars been able to appreciate the diversity of truth claims in world religions. As Wilfred Cantwell Smith has explained, 'It is a surprisingly modern aberration for anyone to think that Christianity is true or that Islam is – since the Enlightenment, basically, when Europe began to postulate religions as intellectualistic systems, patterns of doctrine, so that they could for the first time be labeled "Christianity" and "Buddhism," and could be called true or false.' This relativism enables philosophers to approach the study of religion not simply as a question about beliefs but also as an inquiry into rituals, moral attitudes, and kinds of feeling. Thus a philosopher might posit the existence of a range of defining characteristics, more or fewer of which might be present in the various belief patterns that have been called religions. Each of these defining characteristics would make truth claims within the culture in which it is lodged. Smith, *Questions of Religious Truth* (London 1967), 73, and *The Meaning and End of Religion: A New Approach to Religious Traditions of Mankind* (New York: Macmillan 1963); John H. Hick, *Philosophy of Religion*, 4th ed., (Englewood Cliffs, NJ: Prentice Hall 1990), 110–11.

Given this context, religion can be interpreted as an ideology that transcends a given social order or historical epoch in its essence, although in any given community it will be expressed in the *language* of that place and time. These deeper patterns of thought, whether labelled religions or ideologies, are said to be distinctive in each social order. They will be subject to the usual struggles between dominant and alternative perspectives, and between the elites and the common people. And, although there are undoubted similarities within British, American, and British North American experiences, some scholars now argue that a distinctive 'Canadian' belief pattern is discernible in the three generations following the 1830s.

I would like to thank Dr Murdith McLean for his assistance with these issues.

22 Goldwin French, 'The Evangelical Creed in Canada,' in W.L. Morton, ed., *The Shield of Achilles* (Toronto: McClelland and Stewart 1968), 15–35; Michael Gauvreau, 'Protestantism Transformed: Personal Piety and the Evangelical Social Vision, 1815–1867' in G.A. Rawlyk, ed., *The Canadian Protestant Experience 1760 to 1990* (Burlington: Welch 1990).

23 J. Webster Grant, *A Profusion of Spires: Religion in Nineteenth Century Ontario* (Toronto: University of Toronto Press 1988), 223–4; Westfall, *Two Worlds*; Michael Gauvreau, 'Beyond the Half-Way House: Evangelicalism and the Shaping of English Canadian Culture,' *Acadiensis* (1990), 167–8. In the 1842 enumeration, one in six Canadians said that they did not attend church regularly; in 1871, this figure had dropped to one in 100; Grant, *Profusion of Spires*, cited in Gauvreau, 'Protestantism Transformed,' 58.

So powerful was the evangelical consensus that, despite the profound challenges launched in the 1880s and 1890s by the new Darwinian sciences and by the 'higher criticism' in theological studies – challenges that did have a great impact in Britain and the United States – Canadian Protestant leaders never lost control of these public debates. In battle after intellectual battle, they found accommodations between their revivalism and their concern both for individual salvation and for the intellectual discoveries of the age. As Marguerite Van Die has suggested in her biography of the Methodist leader Nathanael Burwash, 'The new knowledge was integrated selectively, and only in so far as it affirmed the continued cultural and intellectual primacy of the Wesleyan relationship between faith and learning. Believing in the ultimate harmony of reason and religion, Burwash maintained that all new knowledge, if rightly understood, would prove supportive of revealed religion.' Marguerite Van Die, *An Evangelical Mind: Nathanael Burwash and the Methodist Tradition in Canada, 1839–1918* (Montreal: McGill-Queen's University Press 1989); Gauvreau, 'Beyond the Half-Way House,' 173.

24 Richard Allen, *The Social Passion: Religion and Social Reform in Canada 1914–28* (Toronto: University of Toronto Press 1971), 3.

25 Ramsay Cook, *The Regenerators: Social Criticism in Late Victorian English Canada* (Toronto: University of Toronto Press 1985), 196–232.
26 Quoted in Michiel Horn, *The League for Social Reconstruction: Intellectual Origins of the Democratic Left in Canada 1930–1942* (Toronto: University of Toronto Press 1980), 77; for other perspectives, see Allen Mills, *Fool for Christ: The Political Thought of J.S. Woodsworth* (Toronto: University of Toronto Press 1991); Alan Whitehorn, *Canadian Socialism: Essays on the CCF–NDP* (Toronto: Oxford University Press 1992); Joan Sangster, *Dreams of Equality: Women on the Canadian Left, 1920–1950* (Toronto: McClelland and Stewart 1989); and James Naylor, *The New Democracy: Challenging the Social Order in Industrial Ontario, 1914–1925* (Toronto: University of Toronto Press 1991).
27 As the volume that presented these arguments to the Canadian public asserted: 'To those who object that capitalism is "rooted in human nature," we answer: Possibly, but so was cannibalism. We no longer eat each other. A civilization is within our reach in which we shall no longer exploit each other ... The pioneer spirit is not dead, but the frontier which it must conquer has shifted from the physical to the intellectual and spiritual.' Research Committee of the League for Social Reconstruction, *Social Planning for Canada* (Toronto: T. Nelson 1935; Toronto: University of Toronto Press 1975), 69. I would like to thank José Igartua, who permitted me to see his unpublished paper on the debates on Canada's Citizenship Act, 1947, which makes this point – that the CCF then endorsed a bilingual, bicultural Canada.

    Blue-collar, white-collar, and farm families all participated in political reform campaigns. Resistance to the dictates of the capitalist economic order, or at least to those aspects of capitalism that handicapped the farm quest for a just price, became a central theme in prairie farm life. It accompanied the creation of co-ops, the farm men's and women's associations, marketing groups such as the Grain Growers' Grain Company and the Wheat Pools, and, finally, the farmers' political parties. The farmers' successes included significant reforms in three great areas of economic organization: *grain handling*, which was revolutionized by the introduction of government supervision and of farmer-owned, co-operative enterprises; *grain marketing*, which was transformed by various co-operative experiments and later, by the establishment of a government monopoly (the Canadian Wheat Board) over the entire business of wheat exports; and *community economic structures*, which were revised by the introduction of credit unions, co-operative retail stores, and co-op fuel and implement sales. Ian MacPherson, *Each for All: A History of the Co-operative Movement in English Canada, 1900–1945* (Toronto: Macmillan 1979); Jeffery Taylor, *Fashioning Farmers: Ideology, Agricultural Knowledge, and the Manitoba Farm Movement, 1890–1925* (Regina: Canadian Plains Research Centre 1994).

The blue-collar response to similar circumstances was to form unions and cooperatives and to join strikes in the hope of securing a living wage and a more secure existence. These institutions enjoyed varying degrees of acceptance, but in general they commanded the loyalty of a smaller fraction of the workforce in Canada than in Britain and Australia. Within the context of North America, however, their achievements were noteworthy. Business leaders had the same aspirations; Michael Bliss, *A Living Profit: Studies in the Social History of Canadian Business 1883–1911* (Toronto: McClelland and Stewart 1974).

The Canadian labour movement, like the American – and to a greater degree than its British and Australian counterparts – experienced numerous profound internal divisions. The tensions between people of different linguistic and cultural heritages, the misunderstandings arising from differences of gender and skill and region, and the difficulty of establishing personal relations in a vast, newly settled country where work sites were often isolated undoubtedly divided workers one from another. Even within unionized industries, partisans of industry-wide locals fought with people who wished to preserve the craft foundation of certain unions. Moreover, socialists opposed advocates of apolitical, non-partisan, and syndicalist approaches, while French-speaking, Catholic confessional unions rejected secular unions.

Samuel Gompers and the American Federation of Labor (AFL) believed in rewarding political friends and punishing enemies. They also opposed the organization of a union-endorsed Labour or Socialist party. They even rejected government-run welfare and unemployment programs on the grounds that reliance on the government would weaken unions. Gomper's principles were adopted holus-bolus by Canada's Trades and Labor Congress when it accepted AFL domination in 1902. As a result, Canada did not acquire a strong, labour-led social democratic party around the turn of the century, as did Britain, Australia, and certain European nations. Instead, numerous small left-wing movements, including Labour and Marxian and syndicalist groups, competed for public attention. Their struggle against the old-line parties, particularly against the remarkably flexible Liberal party, can be described as unequal at best.

An introduction to this story is contained in John Herd Thompson and Allen Seager, *Canada, 1922–1939: Decades of Discord* (Toronto: McClelland and Stewart 1985); Irving Abella, *Nationalism, Communism and Canadian Labour: The CIO, The Communist Party of Canada and the Canadian Congress of Labour, 1935–1956* (Toronto: University of Toronto Press 1973); Ian Radforth, *Bush Workers and Bosses: Logging in Northern Ontario, 1900–1980* (Toronto: University of Toronto Press 1987); Evelyn Dumas, *The Bitter Thirties in Quebec* (Montreal: Black Rose 1975); Bryan Palmer, *Working-Class Experience: Rethinking the History of Canadian Labour, 1800–1991*, 2nd ed. (Toronto: McClelland and Stewart 1992); Craig Heron, *The Canadian Labour Movement: A Short History* (Toronto: Lorimer 1989); and Desmond Morton with Terry Copp, *Working People: An Illustrated History of Canadian Labour* (Ottawa: Deneau 1980).

Blue-collar political movements, like unions themselves, built slowly between the 1880s and the First World War, exploded in one dramatic confrontation with capital in 1919 (often referred to as the Winnipeg General Strike, but a much wider dissent than this label would suggest), and then split apart over theoretical differences during the 1920s. The events of the next two decades pulled them closer together again. The unionism of the crafts workers, often sectarian but consistently at the centre of the Canadian labour movement, accounted for some of the progress. So too did the entrenchment of cooperative and social democratic ideas in Canada's large farm sector. The economic crisis of the 1930s provided openings for many aggressive Communist union organizers. And the post-1935 resurgence of industry-based union organizing in the United States and Canada injected energy at a crucial moment. Both Communists, who had been freed to consider new strategies by Stalin's endorsation of Popular Fronts in 1935, and socialists, who were determined to reform depression-prone capitalism, could agree on the necessity of organizing drives in the mass production industries such as automobiles, electrical products, and mining.

Having regrouped during the closing years of the Depression, Canadian workers found even greater advantage in the industrial circumstances of the Second World War. Unions grew quickly, workers' militancy kept pace, and in 1943 over 30 per cent of union members participated in strikes. To the dismay of the governing federal Liberal party, the CCF seemed to be making rapid headway in its political organizing. William Lyon Mackenzie King, the Liberal prime minister, responded to these challenges early in 1944 by making unions a recognized partner in industrial relations. By granting them legal protection and guaranteeing a right to collective bargaining, King established the foundations of a system of industrial relations that has continued more or less unchanged to the present. Several years later, unions won the right to levy dues by means of a compulsory check-off, a gain that ensured continuity in union structure and professional organizing personnel.

The social democrats within the union movement also waged a fierce battle to remove Communists from Canadian unions during the 1940s. Despite the costs associated with such an internal struggle, their victory mattered because of its role in the Cold War mapping of political choices: Canadian unions could not be so easily painted by North American business critics as part of an alien system borrowed from the Soviet Union for purposes of subversion.

Chapter 7: Roseanne and Frank Go to Work

1 Roseanne is an invented name. This discussion of her life is based on four interviews that I conducted with her in 1997–8 and on several short, unpublished manuscripts that she has written in recent years. I would like to thank her and her family very much for their generous permission to conduct this inquiry.
2 Ken Dryden, *The Moved and the Shaken: The Story of One Man's Life* (Toronto: Viking 1993). I would like to thank Tom Nesmith for this reference.
3 Ibid., 303.
4 Michael Valpy, 'The War between the Classes,' *Globe and Mail*, citing Ekos Research Associates Inc., *Rethinking Government '94*, based on a year-long study and sample of 2,400 Canadians sixteen years and over, accuracy within 3.2 points 19 times out of 20.
5 Wallace Clement, 'Comparative Class Analysis: Locating Canada in a North American and Nordic Context,' *Canadian Review of Sociology and Anthropology* 27, no. 4 (1990), 462–86. Clement's description of the old middle class included farm, commodity-producing, and

some white-collar households. Among the many works on insecurity, see Richard Sennett, *The Corrosion of Character: The Personal Consequences of Work in the New Capitalism* (New York: W.W. Norton
1999), and Larry Elliott and Dan Atkinson, *The Age of Insecurity*
(London: Verso 1999).

6 June Corman, Meg Luxton, D.W. Livingstone, and Wally Seccombe,
   Recasting Steel Labour: The Stelco Story (Halifax: Fernwood Publishing 1993), 44–53.

7 Reviews of this transition include M.C. Urquhart, 'Canadian Economic Growth 1870–1980,' Queen's University Department of Economics Discussion Paper No. 734 (1988), 50; also Kenneth Norrie
   and Douglas Owram, *A History of the Canadian Economy* (Toronto:
   Harcourt Brace Jovanovich 1991), 553.

8 Canada was part of the advanced industrial world too, because the
   proportion of its population engaged in 'blue-collar' work, in transportation and construction and manufacturing, remained more or
   less stable at about 30 per cent until the 1980s, when a small decline
   occurred.

9 An American agency estimated that 40 per cent of new investment
   in that country in 1988 went to the purchase of information technol-
   ogy – computers, telecommunication devices – as compared to only
   20 per cent in 1978; Robert E. Babe, *Telecommunications in Canada:
   Technology, Industry, and Government* (Toronto: University of Toronto
   Press 1990), 247–8, citing Canada, Department of Communications,
   *Communications for the Twenty-First Century: Media and Messages in
   the Information Age* (Ottawa: Queen's Printer 1987); Herbert I.
   Schiller, *Culture, Inc.: The Corporate Takeover of Public Expression*
   (New York: Oxford University Press 1989), 4, citing the United
   States Office of Technology Assessment from Colin Norman,
   'Rethinking Technology's Role in Economic Change,' *Science* 240 (20
   May 1988), 977; Michael Rogers Rubin and Mary Taylor Huber with
   E.L.Taylor, *The Knowledge Industry in the United States 1960–1980*
   (Princeton, NJ: Princeton University Press 1986).

   The entertainment industry (music, films, magazines, television,
   and video) has been described by filmmaker George Lucas as the
   United States's 'second-largest export industry'; quoted in 'Paramount Victory Would Be Severe Blow to Second-Largest U.S.
   Export Industry,' *Globe and Mail*, 15 July 1989, B4. Lucas noted:
   'Some see advantages for the United States in this kind of foreign
   investment. I, for one, see serious danger in losing our economic

base. For history – from Marco Polo's trade routes to the East India
Company's imperial outposts – teaches that political power is
intrinsically tied to economic power.'

Seven of the world's eight largest design companies were located
in Britain in 1998; these seven account for Can. $30 billion in eco-
nomic activity and employ more than three hundred thousand people; Charles Leadbetter, 'Let Them Eat Weightless Cake,' *New
Statesman and Society* (23 Jan. 1998), 31.

The monetary value of this information and communication
activity is difficult to estimate. See Babe, *Telecommunications*, 23–4;
Paul Audley, *Canada's Cultural Industries: Broadcasting, Publishing,
Records and Film* (Toronto: Lorimer 1983), 5; and William G. Watson,
National Pastimes: The Economics of Canadian Leisure (Vancouver:
Fraser Institute 1988).

The nature of this commodity also required new thinking. Its
value could not be described as a natural product like coal or wheat
because it existed as a few electronic bits on a magnetic tape or disc.
It was not scarce as economics formerly posited scarcity. Mark
Poster, *The Mode of Information: Poststructuralism and Social Context*
(Chicago: University of Chicago Press 1990), 73; Vincent Mosco and
Janet Wasko, eds., *The Political Economy of Information* (Madison:
University of Wisconsin Press 1988); Jim McGuigan, *Cultural Popu-
lism* (London: Routledge 1992); and Nicholas Garnham, *Capitalism
and Communication: Global Culture and the Economics of Information*
(London: Sage 1990).

The challenge to entrepreneurs lay in finding ways of commodi-
fying more of this information and entertainment flow, of charging
per use for access to it, of linking precise monitoring and measuring
with centralized control and efficient distribution. They understood
that there was money to be made in the expanding sector compris-
ing tradeable information and that they would have to ensure its
status as private property. Kevin Robins and Frank Webster, 'Cyber-
netic Capitalism: Information, Technology, Everyday Life,' in
Mosco and Wasko, eds., *Political Economy*, 63.

For examples of the assumptions contained within this profit-
based and privatizing approach, see Jan Fedorowicz, ed., *From
Monopoly to Competition: Telecommunications in Transition* (Missis-
sauga, Ont.: Informatics Publishing 1991), 7.

10 David Harvey, 'From Space to Place and Back Again: Reflections on
the Condition of Postmodernity,' in Jon Bird, Barry Curtis, Tim Put-

nam, George Robertson, and Lisa Tickner, eds., *Mapping the Futures: Local Cultures, Global Change* (London: Routledge 1993), 7.

11 Increasing numbers of women worked for wages. The proportion rose from under 40 per cent of all women aged 25 to 54 in 1966 to nearly 70 per cent in the late 1980s. For comments on the trend, see the Royal Commission on the Economic Union and Development Prospects for Canada, *Report*, vol. II (Ottawa: Queen's Printer 1985), 16; also Alison Prentice et al., *Canadian Women: A History* (Toronto: Harcourt Brace Jovanovich 1988), 367–90; Norrie and Owram, *A History of the Canadian Economy*, 583–4.

12 The increase in the number of jobs in the labour market involved teenagers as well as women. Many of these tasks were less than ideal, so much so that observers began to distinguish between 'good' jobs (the positions that, some mocked, provided a dental plan) and bad ones. Women earned less than their male counterparts, about 15 to 20 per cent less, for similar or equal work, even when they had equal qualifications; see Prentice, *Canadian Women*,   ·   379, citing Ontario, 'Green Paper on Pay Equity – Fact Sheet' (1986). A report by the Economic Council of Canada, *Good Jobs, Bad Jobs: Employment in the Service Economy* (Ottawa: Queen's Printer 1990), argues that the middle-level positions declined in number during the 1980s as the proportion of high- and low-income earners increased. Women figured disproportionately in the secondary labour market, in the so-called bad jobs.

   Most of women's unpaid work still had not been acknowledged by official economic statistics. Calculations of their invisible contributions to the economy reached as high as 40 per cent of a country's production of goods and services, yet none of this labour was evaluated by or included in the national economic accounts; Marilyn Waring, *If Women Counted: A New Feminist Economics* (San Francisco: HarperSanFrancisco 1988, 1990) and an interview with her in *Globe and Mail*, 13 March 1990, A10.

13 Reg Whitaker, *The End of Privacy: How Total Surveillance Is Becoming a Reality* (New York: New Press 1999).

14 Bruce Little, 'Education's Brutal Dividing Line on Jobs,' *Globe and Mail*, 9 March 1998. In the 1990s alone, the unemployment rate among younger workers – aged fifteen to twenty-four – who did not graduate from secondary school was at least double that of graduates from postsecondary programs.

15 Bryan D. Palmer, *Working Class Experience: Rethinking the History of*

Canadian Labour, 1800–1991 (Toronto: McClelland and Stewart 1992), 285, 299, 321–3, 336–9; Craig Heron, *The Canadian Labour Movement: A Short History* (Toronto: Lorimer, 1989); Desmond Morton, 'Strains of Affluence 1945–1987,' in Craig Brown, ed., *The Illustrated History of Canada* (Toronto: Lester and Orpen Dennys 1987), 467–543.

Chapter 8: Culture and Politics Today

1 Among the works that offer valuable advice on these issues, I would like to mention David Harvey, *The Condition of Postmodernity* (Oxford: Basil Blackwell 1989); Ernest Gellner, *Nationalism* (London: Weidenfeld and Nicolson 1997); David Thelen, *Becoming Citizens in the Age of Television* (Chicago: University of Chicago Press 1996); Perry Anderson, *The Origins of Postmodernity* (London: Verso 1998); Joyce Appleby, Lynn Hunt, and Margaret Jacob, *Telling the Truth about History* (New York: W.W. Norton 1994); and the works of Raymond Williams.

2 Ben Pimlott, 'The Politics of Cuddling,' and Charles Nevin, 'Haunted by the Image of Fame,' *Guardian Weekly*, 7 Sept. 1997.

3 Annick Cojean, 'Diana, a Princess with a Big Heart,' *Le Monde*, 27 Aug. 1997, reprinted in *Guardian Weekly*, 7 Sept. 1997.

4 John Lloyd, 'Life without Diana,' *New Statesman* (3 Oct. 1997), 26; *Guardian Weekly*, 'Princess Who Flew Too Close to the Sun,' 7 Sept. 1997.

5 Grant McCracken, *Culture and Consumption: New Approaches to the Symbolic Character of Consumer Goods and Activities* (Bloomington: Indiana University Press 1988, 1990), offers an introduction to these themes.

6 An enthusiastic proponent of this view is John Fiske, in *Television Culture* (London: Methuen 1987) and in *Power Plays/Power Works* (London: Verso 1993). A gentler approach is Ien Ang, *Living Room Wars: Rethinking Media Audiences for a Postmodern World* (London: Routledge 1996).

7 The screen-capitalist age differs from earlier epochs in the degree of importance to be attached to abstract and figurative types of work, the usual acts described as cultural. In this era, as in earlier ones, works of art and intelligence, whether the tract of a philosopher or the conversation of a friend, and daily acts of education, whether the watching of a television show or the teaching of a child, are vital

forces in the establishment (or destruction) of social cohesion and personal meaning. But today symbolic acts occupy a greater proportion of daily life and have greater economic significance. This is not to say that guns and butter do not retain value, only that the design of the gun, the wrapping around the butter, and the movie or advertisement wherein each is featured have taken on a value of their own that is unprecedented in world history. Power has always rested not only with generals and General Motors but with those who create and restate knowledge, who are skilled at conceptualization or at providing articulate commentary, who demonstrate a flair for fashion or for discerning signposts of cultural change, but nowadays these people are more numerous and fulfill a wider range of tasks. Moreover, the cultural and political implications of their work are harder to perceive.

8 Raymond Williams, *Culture* (Glasgow: Fontana 1981), 225, citing P. Bourdieu and J.-C. Passeron, *Reproduction in Education, Society and Culture* (London: Sage 1977). One should distinguish between, for example, the cultural consumption of youth, which is often lucrative but not necessarily of political importance, and the cultural authority of John Maynard Keynes, which has little monetary value in itself but did shape the political discourse of a generation.

9 Paul Rutherford, *The Making of the Canadian Media* (Toronto: McGraw-Hill Ryerson 1978), 78; and Rutherford, *When Television Was Young: Primetime Canada 1952–1967* (Toronto: University of Toronto Press 1990), 12; he writes that Canadians made seventeen cinema visits per year in 1952.

10 Rutherford, *The Making of the Canadian Media*, 80.

11 Mary Vipond, 'Canadian Nationalism and the Plight of Canadian Magazines in the 1920s,' *Canadian Historical Review* 58, no. 1 (March 1977), 43–63; Rutherford, *The Making of the Canadian Media*, 81–2.

12 Rutherford, *When Television Was Young*, 50; he notes (137) that television sets were in 95 per cent of Canadian homes in 1967. Mary Vipond, *Listening In: The First Decade of Canadian Broadcasting, 1922–1932* (Montreal: McGill Queen's University Press 1992); Vipond, *The Mass Media in Canada* (Toronto: Lorimer 1989); Donald G. Wetherell with Irene Kmet, *Useful Pleasures: The Shaping of Leisure in Alberta 1896–1945* (Regina: Canadian Plains Research Centre 1989); and Wayne Schmaltz, *On Air: Radio in Saskatchewan* (Regina: Coteau 1990).

13 Provincial censors regulated these films to ensure that man-woman

embraces on screen were not disturbing and that the language did
not incite revolt; James M. Skinner 'Clean and Decent Movies:
Selected Cases and Responses of the Manitoba Film Censor Board,
1930 to 1950,' *Manitoba History* 14 (1987), 2–9.

14 Gary Evans, *John Grierson and the National Film Board: The Politics of
Wartime Propaganda* (Toronto: University of Toronto Press 1984);
Canada, Royal Commission on National Development in the Arts,
Letters, and Sciences, *Report* (Massey Report) (Ottawa: King's
Printer 1951), 50–9; D.B. Jones, *Movies and Memoranda: An Interpre-
tive History of the National Film Board of Canada* (Ottawa: Canadian
Film Institute 1981); Joyce Nelson, *The Colonized Eye: Rethinking the
Grierson Legend* (Toronto: Between the Lines 1988).

15 Vipond, *Listening In* and *The Mass Media in Canada* (Toronto:
Lorimer 1989). The pattern was reinforced by the passage of a
music content law in 1969; Paul Audley, *Canada's Cultural Industries:
Broadcasting, Publishing, Records and Film* (Toronto: Lorimer 1983).

16 Donald Creighton, 'Watching the Sun Quietly Set on Canada,'
Maclean's (Nov. 1971) 89.

17 Ibid.

18 Paul Rutherford, 'Made in America: The Problem of Mass Culture
in Canada,' in David H. Flaherty and Frank E. Manning, eds., *The
Beaver Bites Back? American Popular Culture in Canada* (Montreal:
McGill-Queen's University Press 1993). See also Mary Jane Miller,
Turn up the Contrast: CBC Television Drama since 1952 (Vancouver:
University of British Columbia Press 1987); Richard Collins, *Culture,
Communication and National Identity: The Case of Canadian Television*
(Toronto: University of Toronto Press 1990); and Rutherford, *When
Television Was Young*.

19 The Massey Commission recognized this; *Report*, 272–3, 277–9; Paul
Litt, *The Muses, The Masses, and the Massey Commission* (Toronto:
University of Toronto Press 1992); J.L. Granatstein, *Canada 1957–
1967: The Years of Uncertainty and Innovation* (Toronto: McClelland
and Stewart, 1986); Bernard Ostry, *The Cultural Connection: An Essay
on Culture and Government Policy in Canada* (Toronto: McClelland
and Stewart 1978).

20 Rutherford, *When Television Was Young*, 13–15, 16, 87.

21 Massey Report, 4, 5, 7, 19–22, 50–9, 271–3, and 277–9. The commissioners described their task as an investigation of those institutions
that 'express national feeling and promote common understanding
and add to the variety and richness of Canadian life.' They were

concerned, they said, with 'human assets, with what might be called in a broad sense spiritual resources.' They argued that 'intangible elements,' things that could not be weighed or measured, had sustained Britain in the Second World War and the French people of Canada during three centuries: 'What may seem unimportant or even irrelevant under the pressure of daily life may well be the thing which endures, which may give a community its power to survive.' The appropriate context in which to think about their concerns is what George Grant described as local and Tory cultural forces; Grant, *Lament for a Nation* (Toronto: McClelland and Stewart 1965).

This discussion finds further amplification in the adult education institutions of the 1930s and 1940s. As Harold Innis wrote in Manitoba Royal Commission on Adult Education, *Report* (Winnipeg 1947), 'A major problem of society emerges in the development of institutions which enlarge the capacities of individuals and enable them to use such enlarged capacities to the greatest advantage of the individual and the institutions.' Reprinted in Harold A. Innis, *The Bias of Communication* (Toronto: University of Toronto Press 1951), 204.

22 Northrop Frye, 'Across the River and Out of the Trees,' in James Polk, ed., *Northrop Frye: Divisions on a Ground: Essays on Canadian Culture* (Toronto: Anansi 1982), 30.

23 John Herd Thompson, 'Canada's Quest for Cultural Sovereignty: Protection, Promotion, and Popular Culture,' in Stephen J. Randall et al., eds., *North America without Borders?: Integrating Canada, the United States, and Mexico* (Calgary: University of Calgary Press, 1992), 269–84; John Meisel, 'Some Canadian Perspectives on Communication Research,' *Canadian Journal of Communication*, special issue (1987), 55–63; R. Bruce Elder, *Image and Identity: Reflections on Canadian Film and Culture* (Waterloo: Wilfrid Laurier University Press 1989). The Canadian-content era coincided with the Trudeau years (1968–79, 1980–4). The Mulroney and Chrétien governments (1984–93 and 1993 on, respectively) have been much more committed to a market-governed cultural policy.

24 Cultural institutions came in all shapes and sizes. Some fulfilled many roles. In addition to the big-budget sites such as movies and national television newscasts, one should note the relevance of local talk radio and the vast network of adult education institutions.

This cautious approach to culture's place in the community can be seen in the outlook of the Massey Commission. The commissioners explained that broadcasting, to use just that one example, was *not* an industry but rather 'a public service directed and controlled in the public interest by a body responsible to Parliament.' They made no bones about this matter: 'We are resolutely opposed to any compromise of the principle on which the system rests and should rest. Radio has been the greatest single factor in creating and fostering a sense of national unity. It has enormous powers to debase and to elevate public understanding and public taste.' And as an 'essential instrument for the promotion of unity and of general education,' the radio system must remain under public regulation. It must not become the creature of 'commercial tendencies,' the Commissioners asserted; that is, radio must not be subject simply to calculations of profit and of the 'lowest common denominator' in public interest. No one could own the airwaves, in this view.

As it was for radio, so it should be for television. Writing about television, this 'new and unpredictable force,' in 1950–1, when it was available in the United States but not in Canada, the commissioners were sufficiently far-sighted to predict that it would be more popular and more persuasive than radio. They also foresaw that, because of the vastly greater investment required by television programming, the pressures for commercial sponsorship would be well-nigh irresistible.

In order to assist Canadian television to avoid the potential traps ahead, they suggested that, given television's close relations with film, the National Film Board should collaborate with and advise the CBC on the purchase and use of films. They argued that the CBC should endeavour to 'import the best programmes from abroad, while developing so far as possible Canadian talent in Canadian programmes.' And they argued that the capital costs of the system should be covered by government grants, while the programming expenses could be met by a special tax on consumers of television, which meant a tax on the purchase of TV sets in this era of the medium's infancy. Thus underlying the commission's approach to television was its conviction that a broadcast medium should be a public service agency and an instrument of national community expression, not an industry to be regulated only by the free market and general standards of decency. If the commission had its way, the state would be a central figure in the development of this latest vehicle of cultural production. Massey Report, 283,

284, 304–5; Collins, *Culture, Communication and National Identity*; *Report of the Task Force on Broadcasting Policy* (Ottawa: Minister of Supply and Services 1986); Frank Peers, *The Public Eye: Television and the Politics of Canadian Broadcasting 1952–1968* (Toronto: University of Toronto Press 1979); Vipond, *Mass Media in Canada*.

25 Roseanne believed emphatically that the pluralism of Canada could be made to work. The Canadian ideal of a national mosaic was much superior, she thought, to the American melting pot. She had heard a great deal about racial tensions in the United States. If Canadian governments could only be persuaded to make fighting racism a priority, she said, the condition of Aboriginal people would quickly improve, but officials had not given the necessary support to these pluralist principles.

26 Some scholars contend that citizens are losing a sense of past and present time. Others believe that they despair of effecting change. On the former, see Eric Hobsbawm, *Age of Extremes: The Short Twentieth Century 1914–1991* (London: Michael Joseph 1994), and George Lipsitz, *Time Passages: Collective Memory and American Popular Culture* (Minneapolis: University of Minnesota Press 1990); on the latter, see Helga Nowotny, *Time: The Modern and Postmodern Experience*, trans. Neville Plaice (1989 [German ed.]; Cambridge: Polity Press 1994).

27 Events since the 1970s have made Donald Creighton's concentration on American dictation of a universal homogeneous culture (the phrase actually comes from his contemporary George Grant) seem less compelling. As cultural studies scholar Ien Ang proposes, the United States might be viewed as leading the transition towards consumption-based cultures that continue to differ substantially in other respects from nation to nation; Ien Ang, *Living Room Wars: Rethinking Media Audiences for a Postmodern World* (London: Routledge 1996), 133–80.

Consumers who lived outside the Canadian mainstream have always read cultural offerings in distinctive ways. Thus the Aboriginal people of northern Manitoba initially perceived the television program 'The Muppets' in terms that would have shocked its producers. Kermit the frog, for example, was a very poor choice of host for the show because of the frog's low place in the Cree and Ojibwa world. The portrayal of the bear, Fozzie, as timid and weak was similarly objectionable, because it flew in the face of Aboriginal respect for a powerful, dangerous, intelligent animal; Gary Granzberg and Christopher Hanks, '"The Muppets" among the

Cree of Manitoba,' *Prairie Forum* 6, no. 2 (1981), 207–10.

The Aboriginal example was perhaps an exception proving the rule. Yet the continued existence of the Canadian political community and of more limited groups – Quebec, Newfoundland, the Inuit, the Nisga'a – suggests that the thesis equating globalization with cultural uniformity is far too sweeping to be convincing.

28 W. Russell Newman, Marion R. Just, Ann N. Crigler, *Common Knowledge: News and the Construction of Political Meaning* (Chicago: University of Chicago Press 1992), 77. I would like to thank Jacques Monet, SJ, for a helpful conversation about his aunt's life.

29 William L Miller, Annis May Timpson, and Michael Lessnoff, *Political Culture in Contemporary Britain: People and Politicians, Principles and Practice* (Oxford: Clarendon Press 1997).

30 Simonne Monet-Chartrand, *Ma vie comme rivière: récit autobiographique*, 4 vols. (Montreal: les éditions du remue-ménage 1981, 1982, 1988, and 1992).

31 Raymond Williams, *Culture* (Glasgow: Fontana 1981), 103–5. It is said that Hollywood film corporations in the 1990s calculated their marketing on the basis of a twelve-week shelf-life for their blockbuster movies. Decline in box office revenues typically commenced after the third or fourth week, well before word of mouth and arm's-length criticism had much effect on audiences' response. Thus, in the case of a film such as *Jurassic Park* or *Titanic*, which had been preceded first by a careful twelve-month campaign, then by a month-long publicity blitz, then by souvenir sales and tie-ins with toy manufacturers and fast-food chains, if the enterprise had not secured a return on investment within the first month, the marketing, not the film itself, must bear much of the blame.

The increasing influence of the market in Canadians' daily lives had one particular effect on their cultural activities. The market's commercial imperatives had become so compelling that its rules of selection and control actually limited the specific works available to the public, especially when large sums of money had been invested. According to Raymond Williams, this was 'the real history of the modern popular newspaper, of the commercial cinema, of the record industry, of art reproduction and, increasingly, of the paperback book. Items within each of these are pre-selected for massive reproduction and, though this may often still fail the general effect is of a relatively formed market, within which the buyer's choice – the original rationale of the market – has been displaced to operate, in majority, within an already selected range.' The actual content and style of the work, and authorial responsibility for it, had thus escaped the market's judgments. Only marketing itself could be evaluated by the profit-and-loss statement.

32 James W. Carey, 'Space, Time, and Communications: A Tribute to Harold Innis,' in his *Communication as Culture: Essays on Media and Society* (New York: Routledge 1992), 167; the ideas are broached in Harold A. Innis, *The Bias of Communication* (Toronto: University of Toronto Press 1951).

33 Carey, 'Space, Time, and Communications,' 167.

34 John Meisel, 'Some Canadian Perspectives on Communication Research,' *Canadian Journal of Communication*, special issue (1987), 58.

35 'Time has been cut into pieces the length of a day's newspaper,' Innis added. These quotations come from Harold Innis, 'Concept of Monopoly and Civilization,' *Explorations* 3 (1954), 89–95, cited in Carey, *Communication as Culture*, 163.

36 I conducted these interviews in February 1998, six months after the events recounted here. I reviewed these statements with her in June 1999, and her opinions had not changed; she was adamant in her insistence on her autonomy and her right to make such decisions. The astonishing 106-week run of Elton John's *Candle in the Wind* on Canada's top-10 singles chart suggests that Roseanne is not alone; Stephanie Nolen, 'Canada Still in Grip of Diana Single,' *Globe and Mail*, 13 Oct. 1999, C1.

37 An illustration from the radio era is William Stott, *Documentary Expression and Thirties America* (New York: Oxford University Press 1973; Chicago: University of Chicago Press 1986).

38 The phrase comes from Pimlott, 'The Politics of Cuddling'; the celebrity system probably began in Hollywood: Jib Fowles, *Starstruck: Celebrity Performers and the American Public* (Washington, DC: Smithsonian Institution Press 1992).

39 Ernest Gellner, *Nationalism* (London: Weidenfeld and Nicolson 1997), and John Gray's review of this book, 'For God and Country,' *New Statesman*, 26 Sept. 1997, 61–2.

40 Eileen Mahoney, 'The Intergovernmental Bureau for Informatics: An International Organization within the Changing World Political Economy,' in Vincent Mosco and Janet Wasko, eds., *The Political Economy of Information* (Madison: University of Wisconsin Press 1988), 297–315; Colleen Roach, 'The U.S. Position on the New World

Information and Communication Order,' *Journal of Communication* 37, no. 4 (1987), 36–51.

41 Lawrence Martin, 'PM's Legacy: Tight Ties to United States,' Montreal *Gazette*, 26 Feb. 1993, B3. The compelling power of such rallying cries as 'Freedom' was matched in contemporary discussion by popular faith in technological progress. One might argue that technology could never be controlled and that markets always made the wisest possible decisions. Andrew Clement, a sociologist at York University, Toronto, does not agree: 'Social organization has an impact on technology before the technology has an impact on social forms. One of the major shaping influences in any technology is the struggle for control – who is going to have a say in how a technology is developed, the ends to which it is put, the way it is used ... The benefits do not accrue to everyone equally in all possible configurations'; Andrew Clement, 'Office Automation and the Technical Control of Information Workers,' in Vincent Mosco and Janet Wasko, eds., *The Political Economy of Information* (Madison: University of Wisconsin Press 1988), 235; this is the message of Robert E. Babe, *Telecommunications in Canada: Technology, Industry, and Government* (Toronto: University of Toronto Press 1990), as well.

Conclusion

1 Recall the statement by Marshall Sahlins cited above in chapter 2: 'The strongest continuity may consist in the logic of the cultural change.' Sahlins, 'Cosmologies of Capitalism: The Trans-Pacific Sector of "The World-System,"' *Proceedings of the British Academy* 74 (1988), 7–9.

2 Jenkins added Roland Barthes's injunction that histories should 'call overt attention to their own processes of production and explicitly indicate the constructed rather than the found nature of their referents.' I have attempted to follow this suggestion; Keith Jenkins, *Re-Thinking History* (London: Routledge 1992), 70.

3 Julie Cruickshank, 'Oral Tradition and Oral History: Reviewing Some Issues,' paper presented to the Canadian Ethnology Association, 1994, 4.

4 A. Irving Hallowell, *The Ojibwa of Berens River, Manitoba: Ethnography into History*, ed. Jennifer S.H. Brown (Fort Worth, Tex.: Harcourt Brace Jovanovich 1992), 60–79.

5 Brian Stock, *The Implications of Literacy: Written Language and Models*

of Interpretation in the Eleventh and Twelfth Centuries (Princeton, NJ: Princeton University Press 1983), 88–91.

6 E.P. Thompson, 'Time, Work-Discipline and Industrial Capitalism,' in his *Customs in Common: Studies in Traditional Popular Culture* (New York: New Press 1993), first pub. in *Past & Present* 38 1967); Eric Hobsbawm, 'Notes on Class Consciousness,' in his *Worlds of Labour: Further Studies in the History of Labour* (London: Weidenfeld and Nicolson 1984).

7 As Janet Spence Fontaine, an Aboriginal colleague of mine on the board of Manitoba Studies in Native History, has written in response to these paragraphs, however: 'We ascribe greater precision to print but know that oral historians were equally precise.' See the discussion in R. Cole Harris, 'Power, Modernity, and Historical Geography,' *Annals of the Association of American Geographers* 81, no. 4 (1991), 672–5.

8 This is the implication of several studies of nationalism, including Benedict Anderson, *Imagined Communities: Reflections on the Origin and Spread of Nationalism* (London: Verso 1991), and Ernest Gellner, *Nationalism* (London: Weidenfeld and Nicolson 1998). A Canadian illustration is presented in Ian McKay, ed., *For a Working-Class Culture in Canada: A Selection of Colin McKay's Writings on Sociology and Political Economy, 1897–1939* (St John's: Canadian Committee on Labour History 1996); McKay often used the phrase 'plain people' to distinguish his constituents; George Bartley also used the term in the *Independent* (Vancouver) c. 1900, cited in Robert A.J. McDonald, *Making Vancouver: Class, Status, and Social Boundaries, 1863–1913* (Vancouver: University of British Columbia Press 1996), 71–3, 116; the phrase was probably used too by the theoretician of the American Populist movement, Charles Macune, as noted in Lawrence Goodwyn, *The Populist Moment: A Short History of the Agrarian Revolt in America* (Oxford: Oxford University Press 1978), 177–9.

9 Walter Ong, *Orality and Literacy: The Technologizing of the Word* (London: Routledge 1982).

10 David Harvey, 'Between Space and Time: Reflections on the Geographic Imagination,' *Annals of the Association of American Geographers* 80, no. 3 (1990), 418–34, and 'From Space to Place and Back Again: Reflections on the Condition of Postmodernity,' in Jon Bird, Barry Curtis, et al., *Mapping the Futures: Local Cultures, Global Change* (London: Routledge 1993), 3–29.

11 George Steiner, *In Bluebeard's Castle*, 66, cited in Keith Jenkins, *Re-Thinking History* (London: Routledge 1991), 63.

12 Eric Hobsbawm, *Age of Extremes: The Short Twentieth Century 1914–1991* (London: Michael Joseph 1994), 16.

13 Charles Tilly, 'Citizenship, Identity and Social History,' in Charles Tilly, ed., *Citizenship, Identity and Social History*, supplement 3, *International Review of Social History* (Cambridge: Cambridge University Press 1996), 1–17; I would like to thank José Igartua for discussions on this subject, as well as this reference.

14 Raymond Williams, *Culture* (Glasgow: Fontana 1981), 13. With these words, Williams offers his definition of 'culture.'

15 Clifford Geertz once addressed the issue of accommodating peoples of sharply different backgrounds within a single community: 'The problem of the integration of cultural life,' he wrote, 'becomes one of making it possible for people inhabiting different worlds to have a genuine, and reciprocal, impact upon one another.' He added that 'if it is true that insofar as there is a general consciousness it consists of the interplay of a disorderly crowd of not wholly commensurable visions, then the vitality of that consciousness depends upon creating the conditions under which such interplay will occur. And for that, the first step is surely to accept the depth of the differences; the second to understand what these differences are; and the third to construct some sort of vocabulary in which they can be publicly formulated.' Clifford Geertz, 'The Way We Think Now: Toward an Ethnography of Modern Thought,' in his *Local Knowledge: Further Essays in Interpretive Anthropology*, (n.p.: Harper Collins Basic Books 1983), 161.

16 Michael Bliss, 'Privatizing the Mind: The Sundering of Canadian History, the Sundering of Canada,' *Journal of Canadian Studies* 26, no. 4 (1991–2), 5–17; J.L. Granatstein, *Who Killed Canadian History?* (Toronto: HarperCollins 1998).

17 Raymond Williams, *Britain in the Sixties: Communications* (Harmondsworth: Penguin Books 1962), 11. The full paragraph reads: 'My own view is that we have been wrong in taking communication as secondary. Many people seem to assume as a matter of course that there is first, reality and then, second, communication about it. We degrade art and learning by supposing that they are always second-hand activities; that there is life, and then afterwards there are these accounts of it. Our commonest political error is the assumption that power – the capacity to govern other men – is the

reality of the whole social process, and so the only context of politics. Our commonest economic error is the assumption that production and trade are our only practical activities, and that they require no other human justification or scrutiny. We need to say what many of us know in experience: that the life of man, and the business of society, cannot be confined to these ends; that the struggle to learn, to describe, to understand, to educate, is a central and necessary part of our humanity. This struggle is not begun, at second hand, after reality has occurred. It is, in itself, a major way in which reality is continually formed and changed. What we call society is not only a network of political and economic arrangements, but also a process of learning and communication.'

18 This line of argument is elaborated in John Ralston Saul, *Reflections of a Siamese Twin: Canada at the End of the Twentieth Century* (Toronto: Viking 1997). I would like to thank him for our stimulating conversations on these matters.

Illustration Credits

Marjorie Goudie: Blake, Baikie, Goudie
Marjorie Goudie and Joe Goudie: Elizabeth, age 16
Rolf Knight: Ali and Phillis Knight; Ali riding the freights; Phyllis and
family dog; Rolf with puppy
Graydon McCrea, Tamarack Films, *Summer of the Loucheux:
Portrait of a Northern Indian Family*: Andre family fishing camp;
Andre family
David Zimmerly: Elizabeth Goudie, age 68

Index

commodification (and consumption), 124, 125, 135, 137, 150–1, 189–92, 193–8, 209, 286n27

communication (*see also* language): through celebrity, 212–14; dominant means of, in society, 5, 7, 31, 149; dreams as medium of, 22–5, 35–6, 38–9, 42–3, 44–7; in Labrador, 59–60, 62–3, 93–4; in oral-traditional society and Aboriginal peoples' politics, 3, 5, 31, 33, 42–7, 54, 239n21; and politics, 95; as primary force in life, 291–2n17; in print-capitalist society, 3, 5, 123, 136–8, 139–51, 264–9nn7–14; through religious language, 82, 87–8, 90; in screen-capitalist society, 3, 5–6, 167, 175, 189, 191–3, 197–8, 202–3, 208–16; through staple exports, 69–70, 75–8, 91; songs and stories as medium of, 24, 32–4, 36, 38–40, 42–3, 48–9, 52, 67, 88, 101, 108, 111–12, 150–1, 204–5, 219–24; technology of, 3, 6, 102–3, 136–8, 140, 149; in textual-settler society, 3, 5, 57–63, 69–70, 75–8, 240–1n1

Communist Party of Canada (CPC), 160–1, 276–7n27

computer, 6, 167, 173, 176–7, 185–6, 215, 223

Conrad, Margaret, 249n30

Conservative Party of Canada, 157, 161

Constitution Act, 17, 133; section 35 of, 19, 243n7

consumption. *See* commodification

continuity, 3, 7, 12–13, 15, 18–19, 31, 33, 44, 47–9, 54, 63–4, 81, 217–20. *See also* Aboriginal people; history

Cook, Ramsay, 231n2, 274n25

Copp, Terry, 261n19, 276

Courville, Serge, 248n26

Craven, Paul, 270n19

Creighton, Donald, 64, 101, 194, 209, 261–2nn22–3, 286n27

Crowley, Terence, 250n10

Cruikshank, Julia, 289n3

Curtis, Bruce, 255n20

culture: as community conversation, 189–90, 192–8, 200–3, 212–15, 222–3, 225, 227; production and consumption of, 189–98, 202–3; reproduction of, 6, 13–16, 32–5, 47, 54, 97, 101, 113–14, 120–1, 132, 151, 225; as tradeable good, 28, 214–15. *See also* history: cultural approach to; markets; Massey Commission

Davis, Angela E., 269n13

democracy, 75, 156–8, 199–203, 209–16; and literacy, 96

Diana, Princess of Wales, 187–9, 211–14

Dick, T.J.O., 262n24

Dickason, Olive, 234nn5, 8

dimensions of time and space, 5–6, 31, 54, 89–91, 218–20, 232n5; have one point of departure when reconstructed, 6; in oral-traditional societies, 12–16, 34–44, 47–54, 57, 218–20, 237n11; in print-capitalist societies, 107, 136–8, 146–8, 162–3, 222–3; in screen-capitalist societies,